COMPUTERS IN
CRITICAL CARE AND
PULMONARY MEDICINE

COMPUTERS IN BIOLOGY AND MEDICINE

Series Editor: George P. Moore
University of Southern California, Los Angeles

COMPUTER ANALYSIS OF NEURONAL STRUCTURES
Edited by Robert D. Lindsay

ANALYSIS OF PHYSIOLOGICAL SYSTEMS
  The White-Noise Approach
Panos Z. Marmarelis and Vasilis Z. Marmarelis

INFORMATION TECHNOLOGY IN HEALTH SCIENCE EDUCATION
Edited by Edward C. DeLand

COMPUTERS IN CRITICAL CARE AND PULMONARY MEDICINE
Edited by Sreedhar Nair

A Continuation Order Plan is available for this series. A continuation order will bring delivery of each new volume immediately upon publication. Volumes are billed only upon actual shipment. For further information please contact the publisher.

COMPUTERS IN CRITICAL CARE AND PULMONARY MEDICINE

Edited by
Sreedhar Nair
Yale University School of Medicine
Norwalk, Connecticut

Associate Editors:

Omar Prakash
Erasmus University, Thoraxcentrum
Rotterdam, The Netherlands

Richard P. Imbruce
University of Bridgeport
Bridgeport, Connecticut

Gary S. Jacobson
Norwalk Hospital
Norwalk, Connecticut

and

Thomas P. Haller
Norwalk Hospital
Norwalk, Connecticut

PLENUM PRESS • NEW YORK AND LONDON

Library of Congress Cataloging in Publication Data

International Symposium on Computers in Critical Care and Pulmonary Medicine, 1st,
  Norwalk Hospital, 1979.
  Computers in critical care and pulmonary medicine.

  (Computers in biology and medicine)
  "Proceedings of the First International Symposium on Computers in Critical Care and
Pulmonary Medicine, held at Norwalk Hospital, Norwalk, Connecticut, May 24–26, 1979."
  Includes index.
  1. Critical care medicine–Data processing–Congresses. 2. Monitoring (Hospital care)–
Data processing–Congresses. 3. Respiration–Data processing–Congresses. I. Nair, Sreedhar.
II. Title. III. Series: Computers in biology and medicine (New York, 1977-   )
[DNLM: 1. Computers–Congresses. 2. Critical care–Congresses. 3. Respiratory system–
Physiology–Congresses. WX218 C7384 1979]
RC86.2.I565 1980   616'.028'0285   80-14503
ISBN 0-306-40449-4

Proceedings of the First International Symposium on
Computers in Critical Care and Pulmonary Medicine, held at
Norwalk Hospital, Norwalk, Connecticut, May 24–26, 1979.

Chairman   Sreedhar Nair, M.D., F.A.C.P.
                Norwalk, Connecticut, U.S.A.

Co-Chairman   Omar Prakash, M.D.
                Rotterdam, The Netherlands

STEERING COMMITTEE

Dr. Richard Imbruce, *Norwalk, U.S.A.*
Dr. Bjorn Jonson, *Lund, Sweden*
Dr. John Osborn, *San Francisco, U.S.A.*
Dr. Richard Peters, *San Diego, U.S.A.*
Dr. Reed Gardner, *Salt Lake City, U.S.A.*
Dr. Byron McLees, *Bethesda, U.S.A.*
Dr. Igal Staw, *Norwalk, U.S.A.*

©1980 Plenum Press, New York
A Division of Plenum Publishing Corporation
227 West 17th Street, New York, N.Y. 10011

Printed in the United States of America

PREFACE

In May 1979 the first international symposium on Computers in Critical Care and Pulmonary Medicine was held at Norwalk Hospital-Yale University School of Medicine. Scientists from eighteen different countries participated in the program which illustrated the importance of computer applications in critical care and respiratory physiology. This book presents the proceedings of the symposium.

I would like to thank Miss Nancy Smith for her untiring efforts and excellent work in typing the manuscripts. Mr. Gary Jacobson and Mr. Thomas Haller have been invaluable in the computerized preparation of the manuscripts and in the use of word processing equipment.

I would also like to gratefully acknowledge the contributions made by my wife Rhoda Nair for her helpful suggestions and her assistance in editing this book.

February 1980   Sreedhar Nair

CONTENTS

ANALOG COMPUTATION FOR EVALUATION OF VENTILATORS

Stanley W. Weitzner, M.D.
Professor, Anesthesiology

Duke University Medical Center
Chief, Anesthesiology Service
V.A. Medical Center
Durham, N.C.

A general purpose, +- 10 volt, solid-state analog computer with electronic mode control of integration (1) has been used since 1966 to calculate, on-line, various parameters of respiratory mechanics. The computer program developed was part of on-going laboratory efforts using dogs for the evaluation of ventilators (2) and, in essence, is an extension of the work of Engstrom and Norlander (3). Figure 1 is a block diagram depicting the experimental setup. Using a Fleisch pneumotachygraph, and appropriate transducers, the following were recorded continuously: airway gas flow, airway pressure, CO_2 concentration at the carina, and central venous pressure (4). The flow, airway pressure, and CO_2 signals were fed to the computer, permitting the calculation, and simultaneous recording, of: tidal volume (VT); expired minute volume ($\dot{V}E$); total dynamic compliance; mean airway pressure; work of ventilation (W-the integral of the product of flow and airway pressure); power of ventilation, which is the rate of work performed -- $\dot{V}$ times airway pressure (the symbol is either Po or $\dot{W}$); and mean expired CO_2 (FEC02). A switching device permitted the recording of these variables on seven recorder channels.

The computer was programmed to solve the standard equations shown in Figure 2. Figure 3 shows the circuit arrangement of the computer modules we used for these studies (5). Inspiratory and expiratory tidal volumes were calculated by integrating the flow signal over a full respiratory cycle, defined as the time between the onset of two consecutive inspirations; during one respiratory cycle therefore, the integrator subtracts from the accumulated inspiratory volume, the expiratory volume. During stable conditions, inspiratory and expiratory tidal volumes were measured as equal, since the volume differences due to the respiratory quotient

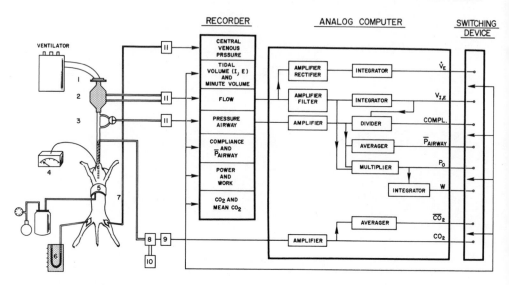

Figure 1: Block Diagram of Experimental Set-up.
1, indicates non-rebreathing valve; 2, pneumotachygraph; 3, location of optional resistor - upper pressure port (P1) was usually used; 4, esophageal ther-
mometer; 5, blood pressure cuff with constant pressure
reservoir; 6, arterial pressure manometer; 7, CVP
catheter; 8 and 9, infrared CO_2 analyzer with output
linearizer; 10, CO_2 sampling pump; and 11, pressure
transducers. For details, see text (Reproduced from
reference 4.)

$$V_I = \int_I^E \left[\frac{dV}{dt}\right] dt \qquad C = \frac{V_I}{P}\bigg|_{\frac{dV}{dt}=0}$$

$$V_E = \int_E^I \left[\frac{dV}{dt}\right] dt \qquad \dot{W} = \frac{dV}{dt}\, P$$

$$\dot{V}_E = \int_0^{60} \left[\frac{dV_E}{dt}\right] dt \qquad W = \int_{I_1}^{I_2} \dot{W} dt$$

$$\bar{P} = \frac{1}{T}\int_0^t P\, dt \qquad \bar{F}_{ECO_2} = \frac{1}{T}\int_0^t F_{CO_2}\, dt$$

Figure 2: Equations the computer was programmed
to solve. For details, see text.

between the breath in and the breath out were too small to be detected with the instrumentation employed. Therefore, this trace
usually starts and ends at the zero baseline (Figure 4).

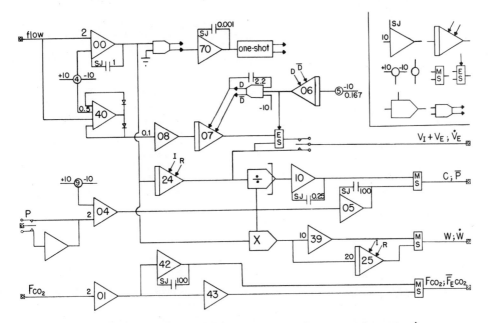

Figure 3: Simplified diagram of the computer circuit. The symbols in the insert (across and then down-
ward) represent: operational amplifier with a gain of
ten (gain is one, when no number is present, and SJ is
a summing junction); integrator, with mode control;
grounded, and ungrounded, potentiometer; motor, and
electronic switch; multiplier, or divider, according
to symbol; comparator with binary output. The
"one-shot" is used to provide integrating (I) or re-set
(R) control mode of operation for integrators 24 and
25. For details, see text and reference 5.

In the initial phase of this study, in which ventilators were evaluated at a constant frequency and I/E ratio, the output of the CO2
analyzer also was fed to an averaging amplifier to derive mean expired CO2. This is possible because inspiratory CO2 concentration
was zero. Although the formula (bottom right in Figure 3) used to
average the expired CO2 is not mathematically correct because it
disregards instantaneous variations in expiratory flow, it was empirically determined (by conventional laboratory techniques) that
this was accurate for the frequency and I/E ratio used (20/minute
and 1:2, respectively).

Figure 4 is a typical trace. The top channel is a central
venous pressure tracing; the second channel down shows tidal volume and the third, airway flow. Airway pressure is recorded on
the fourth channel, with compliance on the fifth channel. Channel

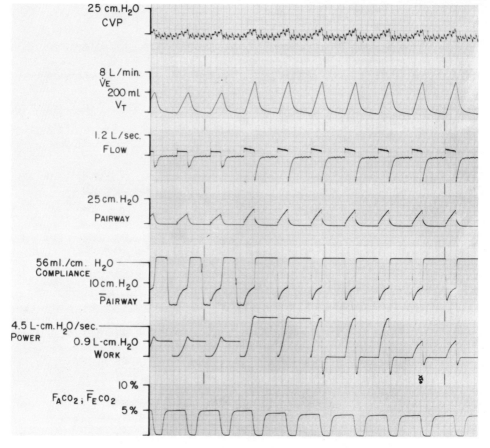

Figure 4:   Record obtained using Celog II Ventilator alone (first three breaths), and with venturi attachment. For details see text.

6 alternately records power and work; the power tracings change
polarity during the two phases of the ventilatory cycle (inspiration is positive) while work always remains positive, because the
work done during inspiration is greater than the (passive) work of
exhalation.

This tracing was obtained using a Celog ventilator (4), in a
paralyzed anesthetized dog. This ventilator is a minute volume divider with a built-in I/E ratio of 1:2. The frequency was set at
20 and the minute volume adjusted so that expired CO2 was 5 percent. Between the third and fourth breaths on the tracing a venturi

mechanism placed between the outlet of the ventilator and the tracheal tube was turned on; this increased volume approximately 50
percent (i.e., 50 percent dilution by the venturi).   Peak flow increases and VT increases 50 percent -- although flow does not maintain a square wave shape. Airway pressure increases correspondingly. The work and power are greatly increased; expired CO2 decreases as a result of the increased ventilation. It should be noted
that total compliance (channel 5), which was obtained be continuously dividing volume by airway pressure, and then measuring only
that point on the recording which occurred when flow was zero at
the end of inspiration, did not change. This point, which is representative of the total dynamic compliance, is conveniently
marked by a "knee" in the curve which remains constant at
21ml/cmH2O (3 large boxes above zero).

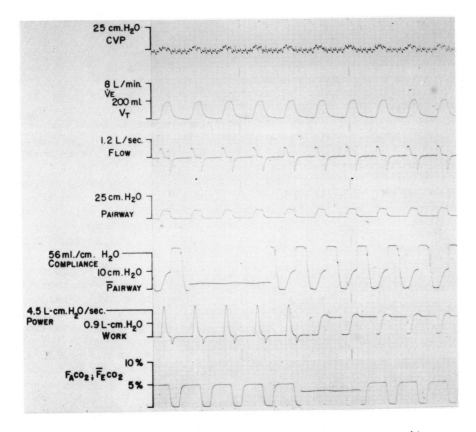

Figure 5: Record obtained using Celog II Ventilator with "bag-in-box" attachment.   For details, see
text (Reproduced from reference 4.)

Figure 5 is a record obtained using the Celog II and its commercially supplied option -- a "bag-in-the box" attachment (5). With this option, a separate inspiratory gas source is used to fill the bag (which is then emptied by the ventilator). Although the Celog II still delivers a low-flow, square wave pattern, for this recording the volume delivered to the box is approximately twice the minute volume. The peak flow is increased and the inspiratory flow time is shorter (compare with Figure 4), while the inspiratory to expiratory ratio remains 1:2. Thus the bag is emptied sooner and there is a period of no-flow with maintenance of airway pressure at the end of inspiration. This is also reflected in the tracings of VT, power, and work.

An electronic switch and a motor-driven multi-ganged double-pole single-throw switch (ES and MS -- also Figure 3) permit respectively, the additional recording of minute volume on channel 2, mean airway pressure (channel 5), and $\overline{FECO_2}$ (channel 7) for 10 seconds out of every 60. This motor switch also permits the alternate recording, for 30 seconds each, of work and power.

The initial experimental protocol was set up as follows. Following an equilibration (control) period in which the ventilator was adjusted to provide a CO_2 concentration of 5 percent (and after which it was not readjusted), the anesthetized and paralyzed dog's chest was compressed at constant pressure (Figure 1) for 15 minutes by a blood pressure cuff inflated to 40 torr (to simulate decreased compliance). Following a ten minute "recovery" period, a unidirectional (inspiratory only) 2mm i.d. resistor was inserted in the tracheal tube adaptor for 10 minutes (to simulate increased airway resistance) and then removed, after which there was a second 10 minute "recovery" period. At the end of each of these periods, and in the middle of chest cuff compression, esophageal temperature and arterial blood gases were measured. Changes in blood gases, VD/VT ratios and $\dot{V}A$ were analyzed, along with the data obtained from the recordings, in an attempt to uncover physiological advantages to a particular ventilator. Two different ventilators were used (in random order) in each dog; in all, over twenty ventilators were evaluated using this or similar protocols between 1966 and 1976 (4,5,6,7). There were no significant differences found when comparing flow or pressure pattern versus arterial O_2 tension. There were, of course, differences among ventilators when considering work capability, ease of control, stability and linearity of control settings, etc.

A typical trace from this type of study is shown in Figure 6 (note the change in the calibration for power and work). Between channels the paper is marked every hundred mm. The beginning and last portions of this strip are run at 10 mm/second paper speed; just after the beginning of the slowly recorded portion in between the two, constant pressure chest compression is begun. One can see increases in CVP, airway pressure, power, and work of ventilation.

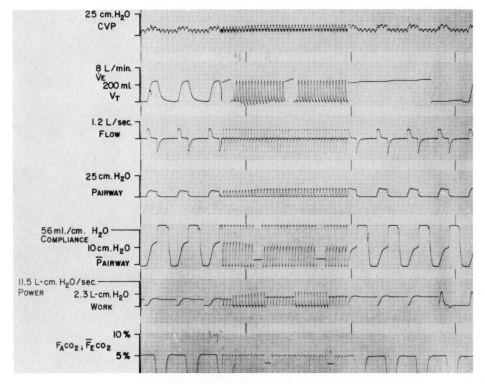

Figure 6: Record obtained using the Celog II Ventilator with "bag-in-box" attachment during constant pressure chest compression. Compression begins just before the first inter-channel dark vertical marker. For details see text. (Reproduced from reference 5.)

At the same time, compliance (the knee of the curve recorded at 10 mm/second, and the top of the darker lower portion of the tracing made at slow speed) is decreased from 35 to 28 ml/cm H_2O. The tidal volume tracing (channel 2) on the down stroke (expiratory tidal volume) is greater than that of the inspiratory tidal volume. (An enlargement of this channel trace is seen in Figure 6A.) This is a reflection of a decrease in total lung volume, due to compression of the soft chest wall of the dog. This study was done using a Celog II ventilator with bag-in-box attachment. The constant-volume ventilator automatically raised airway pressure in order to maintain constant minute and tidal volumes. CO_2 falls slightly (5.2 to 4.9 percent); the mechanism for this fall in CO_2 can be deduced from the tracing -- for the first nine breaths following chest compression (Figure 6A) the volume out was greater than the volume in, indicating a smaller lung being ventilated at constant volumes. This relative hyperventilation caused a decrease in expired CO_2 (2,5).

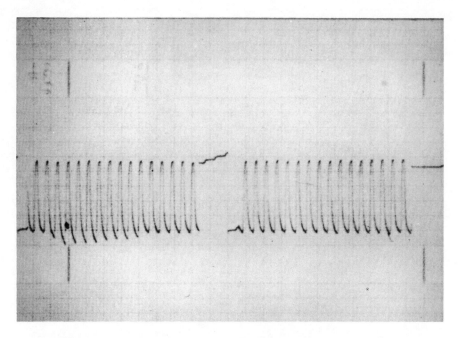

Figure 6A: Enlargement of channel 2 from figure
6. Tidal volume out goes below the baseline and is
therefore greater than the inspiratory tidal volume.
Note also the record of $\dot{V}E$ in the center of the figure.
For details, see text.

Figure 7 is a composite of three periods of recording from the
same paralyzed dog. The left-hand third is the control tracing and
shows a VT of approximately 250 ml being delivered. In the fourth
channel down, the "knee" on the compliance curve is indicated by a
small arrow (pointing down and to the right); this part of the
continuously recorded curve occurs at the end of inspiration when
flow is zero, and marks the point at which we measure dynamic compliance. The middle third shows the same dog after the inspiratory
resistance was put in his airway (see number 3 Figure 1) Airway
pressure is measured at the upper port (Pl), which is between the
inspiratory resistor and the pneumotachygraph. In this middle
panel the calibration mark indicates 30 cm H2O, rather than 15 cm
H2O airway pressure, and the compliance and mean airway pressure
calibrations are attenuated by one-half. With the resistor in
place, and a higher airway pressure, volume delivered by the Engstrom Ventilator is decreased due to the internal compressible dead
space of the ventilator and tubing. Inspiratory flow lasts
throughout the entire period of the inspiratory phase time; the
work and power delivered by the ventilator are significantly increase-

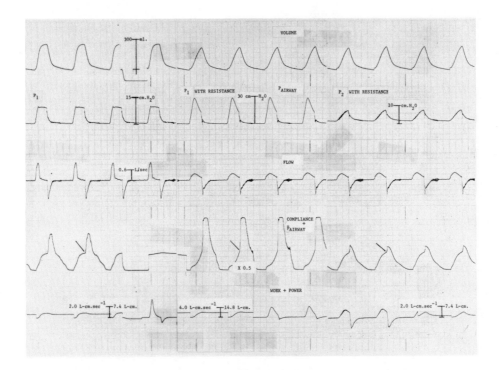

Figure 7: Composite of tracings obtained using an Engstrom Ventilator in an anesthetized, paralyzed, dog (first third is control period). The middle third of the tracing was obtained after inserting a resistor in the tracheal tube and measuring airway pressure at port P1 (see figure 1). The last third was obtained with the resistor in place, but measuring airway pressure through port P2. For details, see text.

ased (note also the change in calibration of the bottom channel). The third panel shows the same parameters recorded using pressure port P2 (located between the inspiratory resistor and the dog). The scales and calibrations are the same in this panel as they were in the control period. The compliance indicated on this portion is identical to that of control. This is as expected, since the inspiratory resistor should affect resistance to flow (i.e., machine airway pressure, and flow pattern) but not compliance.

Figure 8 was recorded from an anesthetized paralyzed dog that had had constant ventilation for over an hour (Drager Spiromat 661 Ventilator) adjusted to give an expired CO_2 of 5 percent. This recording shows most clearly the ability of the electronic and the

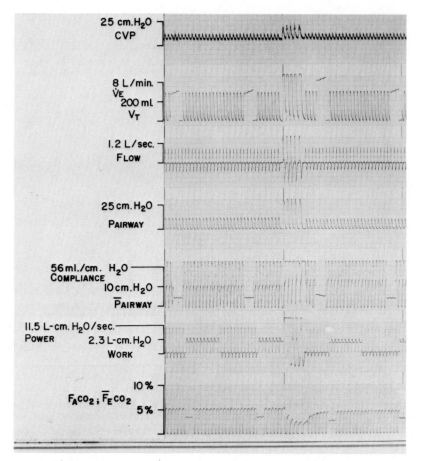

Figure 8: Tracing obtained in a mechanically ventilated dog showing the effect of 5 manual breaths
("deep sigh"). See text for details. (Reproduced from
reference 4).

mechanical switches to permit us to record -- for ten seconds out
of every sixty -- two different variables (note second, fifth, and
seventh channel from the top in which, respectively, $\dot{V}E$, mean airway pressure, and $\overline{F}ECO2$ are recorded). The Drager Ventilator was
chosen to investigate the equivalent of a "deep sigh" because it
permitted the operator to manually administer deep breaths without
changing any of the ventilator controls, then allowing resumption
of the same artifical ventilation. Five deep breaths are recorded
in the middle of the tracing. (An enlargement of the second channel (VT, $\dot{V}E$) is seen in Figure 8A.) CVP, VT, inspiratory and expi-

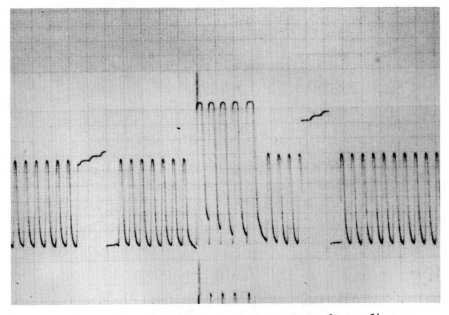

Figure 8A: Enlargement of channel 2 from figure
8:   the effects of "deep sighs." Note that trace of VT
does not return to baseline. VT out is less than inspiratory VT, indicating an increase of functional residual capacity. See text for details.

ratory flow, and airway pressure were more than doubled, while $\dot{V}E$,
for that minute, was increased about 50 percent; CO_2 concentration
consequently was markedly reduced. Note that the volume out during
these five breaths is less than the volume in (Figure 8A) -- indicating an increase in functional residual capacity. With equilibrium nearly re-established about sixty seconds later, compliance
is greater than it had been before, reflecting the altered condition of the lung. When using constant-volume ventilators airway
pressure is primarily dependent on the condition of the lung;
therefore it is decreased after the deep breaths. Minute volume is
actually higher than it had been before because each delivered
tidal volume is increased. As a result of the improved compliance,
airway pressure is lower and there is less loss to the internal
compliance of the ventilator. In addition, the power and work
needed by the ventilator are decreased. With the decrease in airway pressure, and the presence of constant volume, work is decresed
-- characteristic of a volume pre-set ventilator. As the result of
the increased ventilation, expired CO_2 falls from 5 to 4 1/2 percent.

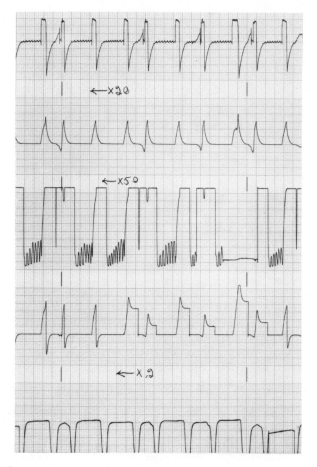

Figure 9: Tracing obtained in an anesthetized, spontaneously breathing dog. A British Oxygen Corporation Harlow Ventilator, in the assist mode, was used. The second, fifth, seventh, and ninth breaths were triggered by the dog. See text for details.

Figure 9 is a tracing taken in an anesthestized dog, unparalyzed, being ventilated by a Harlow Ventilator (very similar in operation to a Bird Mark VII -- pressure pre-set, variable flow, with expiratory time and trigger sensitivity controls). Channel 1 (top channel) is flow at the airway, channel 2 is airway pressure (going negative when the dog takes a spontaneous breath in). Channel 3 is compliance, with a ten-second strip of mean airway pressure; recordings on this channel are off scale because pressure and flow are set at high sensitivities so that the work and power tracings in channel 4 are greatly expanded. The bottom chan-

nel is carinal CO_2 (the lower portion of which is cut off). The
second breath on the record is initiated by the dog's spontaneous
effort and shows a negative airway pressure (with some resulting
inspiratory flow) immediately triggering a machine breath (note the
difference in flow pattern in top channel). Since this occurs
sooner than a regularly programmed (timed by the ventilator) tidal
volume, the CO_2 for that breath is lower than it is for a machine
breath. The first three breaths on channel 4 show inspiratory and
expiratory power. The remainder of this tracing until the very
last breath records inspiratory and expiratory work and in the
fifth, seventh and ninth breaths, the work of "triggering" the ventilator. With a triggered breath there is a negative-going value
for work immediately preceding. This represents the work the dog
has to do in order to trigger the ventilator. This record was made
during a phase of studying "patient-trigger" (assistor) mechanisms
of several pressure-cycled and volume pre-set ventilators.

Description of a trigger mechanism, in order to be complete,
should include the inspiratory work that must be done by the patient, and the internal time delay (response time) between activating
the trigger mechanism's sensing device and initiation of ventilator
inspiratory flow. Inspiratory work is a function of the volume of
gas the patient must move in order to generate the negative pressure needed to trigger, and of how long he must maintain this
(i.e., work equals the integral, with respect to time, of airway
pressure times flow, during the period of inspiratory effort; or,
work equals the integral of (P[airway] times $\dot{V}$ dt). The time limits of this integration may be affected also by the response time
of the trigger sensor itself, which is different from the internal
time delay mentioned above. This response time is extremely small
and usually of no clinical consequence. Implicit in the formula
just listed is the possibility that the work a patient must do in
order to trigger a ventilator may vary from apparatus to apparatus
even though each one be set to trigger at the same negative inspiratory pressure. For example, the measured work required to
trigger the Harlow pressure-cycled ventilator is some six times
greater than that required to trigger the Bird Mark VII, even
though each ventilator is set to assist at the same negative pressure. The reason for this becomes apparent when one considers
where the trigger sensor is located in relation to the patient's
airway in each ventilator. In the Harlow, the sensor is deep within the ventilator itself; the dog had to do more work in order to
transmit the same negative pressure through a larger volume. In
addition, the trigger response time is longer in the Harlow, increasing work significantly.

This is more clearly illustrated by the tracing in Figure 10,
run at a paper speed of 25 mm/sec. The top channel is tidal volume, the second channel down is airway flow, the third channel is
airway pressure, the fourth channel is compliance, and fifth channel, work. There is only a single breath depicted on this tracing,

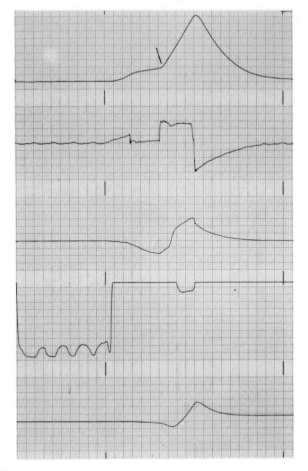

Figure 10: Tracing of a single triggered breath
(5 large boxes equal one second) taken from the same
study as figure 9. Note the delay between the start of
negative airway pressure (channel 3) and the initiation
of the machine breath (rapid rise of "square wave" flow
in channel 2). See text for details.

which is taken from the same study as Figure 9. Again, most of the
continous recording of compliance is off scale because pressure and
flow were measured at calibrations sensitive enough to record inspiratory work (bottom channel). In the third channel down, ten
large boxes from the left margin, one can see airway pressure begin
to leave zero as the dog begins an inspiratory effort; there is a
corresponding rise in the airway flow (channel 2), and as a result
a small inspired tidal volume (the small first peak). From that
point to the <u>beginning</u> of the triggered ventilator breath (indicat-

ed by the diagonal line pointing from left to right), there is a
640 millisecond delay, during which the dog continues his inspiratory effort against a closed valve (causing airway pressure to continue to increase negatively). The amount of work this requires is
indicated (negative deflection) in the bottom channel. In the VT
tracing, the volume delivered by the machine is represented by the
distance between the knee and the top of the curve. The use of the
analog computer, in real time, facilitated this breath-by-breath
measurement and analysis in vivo.

Figure 11 is taken from a dog being ventilated by an Emerson
Postoperative Ventilator adapted with an IMV kit provided by the
Emerson Company. The top channel is expired CO_2, the second is
CVP, the third channel is airway flow, and the bottom channel is
airway pressure. We attempted to adjust the ventilator volume so
that expired CO_2 was 5 percent and then slowly decreased the rate
control until there were mandatory machine breaths 4 times a minute. The first part of this tracing is run at a paper speed of 2
1/2 mm/second and the last portion at 1 mm/second. The difference
between spontaneous and machine breaths is obvious, and the ability

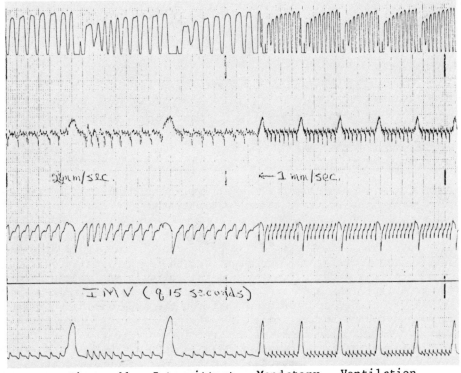

Figure 11: Intermittent Mandatory Ventilation
(f = 4/minute) in a dog using Emerson Postoperative
Ventilator. See text for details.

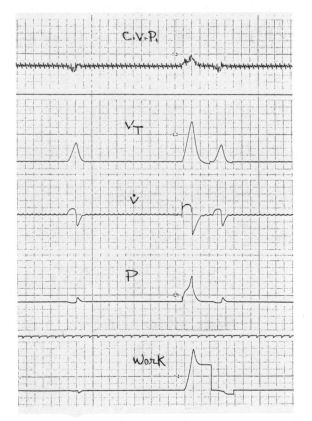

Figure 12. Intermittent Mandatory Ventilation
(f = 2-3/minute) in a dog using Bird IMV Ventilator.
See text for details.

of this mode of ventilation to keep expired carbon dioxide at, or below, a desired value is clearly illustrated.

Figure 12 is a photograph of a tracing taken from a spontaneously breathing anesthetized dog being ventilated with a Bird Corporation IMV Ventilator. The first breath is a spontaneous one, the second is a machine breath, and the third is again spontaneous. This ventilator has a demand system which is very sensitive, permitting spontaneous breathing throught the body of the ventilator (it does not have any external IMV valving, reservoir, and/or circuitry); in addition there is a small internal reservoir to permit additional volume to be taken on demand should the demand valve flow not be sufficient. The time markings (recorded between the fourth (airway pressure) and bottom channels) indicate one second intervals. The bottom channel shows a record of power on the first breath, with work of ventilation recorded during the second and third breaths. The minimal amount of work needed for spontaneous breathing with this ventilator is clearly depicted during the third breath.

This computer has been used over the past twelve to thirteen years to study evolving intensive respiratory care unit therapy and changing ventilatory maneuvers used in clinical practice; it has been used to contrast peep and expiratory retard, and to study IMV and synchronized IMV, the mechanics and physiological effects of servo-ventilator ventilation, as well as the application of single and multiple-unit fluidic ventilators (7). An analog computer was chosen because of its high-speed computing ability, its relatively small expense, its ease of programming and reprogramming, and the ability, together with the recording format, to present continous analog signals which we physicians were already trained to interpret. The importance of the ability to modify a study by having results available on-line are obvious and well documented. The system was easy to set up and calibrate, taking only one-half hour for computer, transducers, and recorder. Its versatility, accuracy, and speed fill an obvious need, and its ability to display copious data in the formats depicted in the illustrations, so as to facilitate rapid, and almost intuitive, appreciation of complex relationships, is clear-cut.

Although this particular computer is not even produced any longer, it is still used for medical student and house staff instruction, serving admirably for this purpose; it had been briefly used for patient monitoring (although its relatively large bulk made it awkward for this application), and most recently was employed in physiological studies of sodium salicylate - induced ventilatory failure (8). More modern technology would permit reduction in its size, and the addition of digital logic and microcomputer versatility would permit even greater application (9) of this system both in the laboratory and in the clinical setting.

REFERENCES

1. Electronic Associates, Inc. Long Branch N.J. - TR-48 Analog
    Computer Reference Handbook - Publ. D800 2008 0A, 1964.

2. AMAHA, K., LIU, P., WEITZNER, S.W. and HARMEL, M.H.: Effect
    of Constant Chest Compression on the Mechanical and Phy-
    siological   Performance   of   Different   Ventilators.
    Anesthesiology 28:498, 1967.

3. ENGSTROM, C.G., and NORLANDER, O.P.: A New Method for the Ana-
    lysis of Respiratory Work by Measurements of the Actual
    Power as a Function of Gas Flow, Pressure, and Time.
    Acta Anaesthesiol. Scand. 6:49, 1962.

4. WEITZNER, S.W. and URBAN, B.J.: A New Ventilator Utilizing
    Fluid Logic. J.A.M.A. 207:1126-1130, 1969.

5. URBAN, B.J. and WEITZNER, S.W.: Analogue Computation for Res-
    piratory   Mechanics.   Progress   in   Anesthesiology:
    Proceedings Fourth World Congress of Anaesthesiologists.
    Edited by Boulton, T.B., et al., Excerpta Medica Founda-
    tion, Amsterdam, 1970. pp 611-617.

6. WEITZNER, S.W. and URBAN, B.J.: Fluid Amplifiers: A New Ap-
    proach to the Construction of Ventilators. ibid. pg.
    1068-1074.

7. URBAN, B.J. and WEITZNER, S.W.: The Amsterdam Infant Ventila-
    tor and the Ayre T-Piece in Mechanical Ventilation.
    Anesthesiology 40:423-432, 1974.

8. LEVINE, S.: Ventilatory Stimulation by Sodium Salicylate:
    Role of Thoracic Receptors. J. Appl. Physiol.
    41:639-645, 1976.

9. LINNARSSON, D. and LINDBORG, B.: Breath-by-Breath Measurement
    of Respiratory Gas Exchange Using On-Line Analog Computa-
    tion. Scan. J. Clin. Lab. Invest. 34:219-224, 1974.

VENTILATOR SURVEILLANCE - ROUTINE APPLICATION AND QUALITY CONTROL

Christen C. Rattenborg, Robert Buccini,
John Kestner, Ray Mikula

The University of Chicago
Department of Anesthesiology
Chicago, Illinois

A ventilator surveillance system has been designed to allow better monitoring of the care of ventilator patients outside intensive care areas. The main reasons for administration of artificial ventilation outside intensive care units have been the fear of superinfection in lymphoma patients, the prolonged artificial ventilation needed for some in neurosurgical patients, and the problem acute overload of intensive care areas.

The patient's airway pressure supplies reliable information about many aspects of the mechanics of ventilation and it is sensitive to most ventilator malfunctions and emergency situations.

METHOD

During the past year, we have continuously monitored all ventilator patients outside the ICU by using bedside telephones for data transmission to a central facility . The airway pressure is obtained from a port of the expiratory circuit and coupled to a pressure transmitter which was built around a National Semiconductor LX 3701D pressure transducer. The output of the pressure transducer controls the frequency of an Intersil 8038 function generator, which is acoustically coupled to the telephone. At the central monitoring station, the signal is demodulated by means of a National Semiconductor phase-locked loop No. 565. There are no external adjustments, and any transmitter is compatible with any of the receiving channels. The signal is displayed continuously on a strip chart recorder. For ease and accuracy of operation, the baseline and gain of the system are precalibrated: 1 mm deflection corresponds to 1 cm H20. The normal recording speed is 25 mm/min.

19

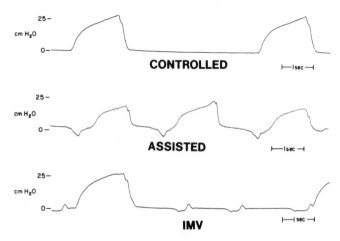

Figure 1.   The common ventilation patterns,   which are observed during mechanical ventilation are:

1.   Controlled ventilation, no patient   effort   is   observed during the whole cycle.

2.   Assisted   ventilation,   a   sub-atmospheric   airway pressure   preceeds   the   inflation, the inflation curve shows a "hump" which is typical for   assisted   ventilation.

3.   IMV means intermittent mandatory ventilation.   The controlled   inflation is supplied at a fixed rate.   The patient breathes spontaneously in the intervals between the inflations.

For   detailed analysis, the recorder is automatically operated at a high speed of 25 mm/sec for 10 sec once every hour.

At the start of surveillance, the transducer is   pneumatically connected to the expiratory tubing, and to the electric power.   The bedside telephone is the link to the   central   switchboard   of   the monitoring system.

Most contemporary ventilators supply a square-flow   inflation; therefore the airway pressure curve is ideally suited to monitoring of resistance, compliance, and work of breathing.   In   a   previous report,   we   have identified normal and abnormal patterns of airway pressure, including various emergencies.[2]

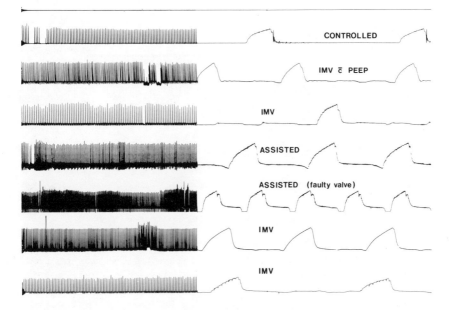

Figure 2. The left side shows the continous slow
movement of the paper, fast movement is seen to the
right.

Channel 1: On standby.

Channel 2: Controlled ventilation.

Channel 3: Servo-assisted IMV (Bear Ventilator). PEEP
is observed as a drop in pressure during suctioning.

Channel 4: Controlled ventilation with IMV (MA-1).

Channel 5: Assisted ventilation

Channel 6: Assisted ventilation with low trigger sensitivity.

Channel 7: Servo-assisted IMV.

Channel 8: Controlled IMV (Bennett MA-1)

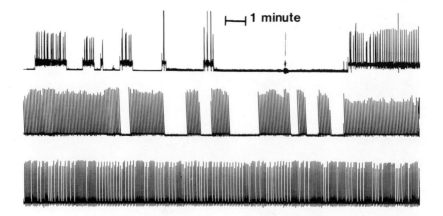

LONG AND SHORT DISCONNECTIONS

Figure 3. A series of long and short interruptions is observed, the long interruptions are up to several minutes indicate periods during which bag ventilation is used in order to mobilize secretions. The short interruptions, lasting 15-20 seconds reflect normal tracheal suctioning. Short interruptions are also shown in Figures 2 and Figure 4.

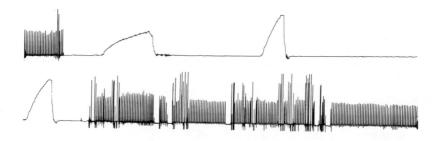

AIRWAY OBSTRUCTION

Figure 4. The high pressure curve with a flat peak is caused by the pressure-limiting valve. Apparently the curves, which are related to suction periods reflect cough.

WATER IN INSPIRATORY LINE

WATER IN EXPIRATORY LINE

Figure 5. Water in the inspiratory tube causes an
irregular ripple during the rising take out pressure.
Expiratory water causes fluctuations when the flow is
high, early in the expiratory phase.

APPLICATIONS

We are here concerned virtually with the quality control of
routine nursing procedures; however, the surveillance system has
also been used for retrospective audit of all phases of ventilator
utilization.

RESULTS

Information was gathered from analysis of the 1000 hours of
artificial ventilation outside ICU areas. The figure legends
(Figs. 1-5) illustrate the application of the system during normal
and abnormal patterns of mechanical ventilation.

Two features used for quality control of nursing care were easily observed from the curves:(1) A series of short interruptions,
lasting 15-20 seconds, invariably indicate suctioning. (2)Ripples
in the inspiratory or expiratory pressure indicate the presence of
water in the corrugated tubing. The short interruptions demonstrate how closely routine suctioning techniques are adhered to.
The response to water in the tubing is more complex: It requires

1. knowing that water in the tubes is abnormal and that it
    should be removed.

2. that the water be observed.

3. knowing how to remove the water, or how to call for assis-
    tance;

4. actual removal of the water (see Figure 6).

INSPIRATORY WATER (REMOVED)

Figure 6. Inspiratory water is evident from the fine ripples; also, the peak pressure is irregular. At the arrow, the peak pressure becomes constant, and the next fast curve shows that the water has disappeared.

REPRESENTATIVE PATTERNS

Pressure curves were chosen from recordings made during the 1000 hours of ventilator surveillance.

The most common patterns of assisted breathing, controlled ventilation, and IMV are demonstrated in Figures 1 and 2.

Longer interruptions (Fig. 3) can have many causes; generally, the patient is ventilated by bag. Short interruptions lasting 15-20 seconds, indicate endotracheal suctioning.

The intermittent high pressures seen in (Fig. 4) were in some cases, limited by a flat peak, indicating activation of the pressure-limiting valve of the ventilator.

Water accumulated in the tubes cause characteristic change (Fig.5). We do not try to make a distinction between the presence of water in the patient airways and in the ventilator tubing. The emptying of water is shown in Fig. 6.

QUALITY CONTROL - SUCTIONING PROCEDURES AND WATER IN TUBES

Proper nursing care depends on many factors, some of which are related to a standard routine. Others depend on detection of problems, followed by proper action to correct them. The occurrence of water in the tubing is an example. Inspiratory water was present early during all nursing shifts, but less so towards the end. Expiratory water had a more uniform time distribution, but was most frequent in the afternoon.

The frequency of suctioning followed a pattern (Fig. 7) which was closely related to nursing shifts and routine, with each nursing shift having its characteristic pattern.

Frequency of Specific Patterns Vs. Time of Day

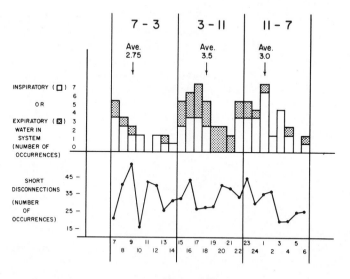

Hour of Day

Figure 7. The events are arranged according to hours and according to nursing shifts. Inspiratory water was present most frequently near the beginning of each shift, whereas the presence of expiratory water was distributed more evenly over time. Some water was observed during each 11-hour period. It is difficult to account for the difference in accumulation of water. It might be related to ventilator rounds by respiratory therapists.

The morning shift had a period of very active suctioning, followed by low activity. During the evening shift, the suctioning was most regular. During the night, there was regular activity until 2:00 AM, after which it slowed down, reflecting the need of the patients to get some uninterrupted sleep.

CONCLUSION

Remote monitoring of ventilator function by telemetry is well suited for identification of various ventilatory patterns. Short interruptions in ventilation provide information about adherence to established practices. For endotracheal suctioning, accumulation and emptying of water from tubing reflects a complex response: awareness of the water, knowledge that the presence of water is wrong, and a decision to remove a decision to remove the water and the knowledge of how to do so. Good nursing care is sine qua non for ventilator patients and care is a sine qua non for ventilator patients, and the routine described constitutes the most important aspects of nursing care.

REFERENCES

1 RATTENBORG, C.C., HOLADAY, D.A.: Constant flow inflation of the lungs. Acta Anaesthesiologica Scandinavicae Proceedings 1 of the Second European Congress of Anesthesiology, Copenhagen. Iussu Societatis Anaesthesiologicae Scandinavicae Edita Supplementum XXIII: 211-213, 1966.

2 RATTENBORG, C.C., MIKULA, R.J.: Ventilator surveillance. Critical Care Medicine Vol 5, p. 252, 1977.

AUTOMATED ESTIMATION OF RESPIRATORY DEAD SPACE: TIDAL VOLUME RATIO

Frank J. Wilson, Jr., M.D.,   James. P. Florez, M.D.,
Roger C. Bone M.D.,   F. Charles Hiller, M.D.

Pulmonary Division
University of Arkansas for Medical Sciences
Little Rock, Arkansas

Dead space:tidal volume ratio (VD/VT) is increased in most pulmonary disorders (1,2,3,). The simplicity of the measurements and calculations required to obtain VD/VT makes it attractive for monitoring changes in lung function in critically ill patients. The usual procedure for obtaining VD/VT is to collect expired gas in a bag for several minutes and to analyze the contents to obtain mixed expired carbon dioxide tension (PECO2). A simultaneous measurement of arterial carbon dioxide tension (PACO2) is made, and VD/VT is calculated manually using the Enghoff modification of the Bohr equation. The effort required to collect mixed expired gas and the necessity for arterial blood sampling make it impractical to determine VD/VT in this manner frequently enough for it to be used for routine monitoring.

We have devised a method for estimating VD/VT automatically using only measurements of airway and expired gas concentrations. The necessary gas analysis makes automatic calculation of respiratory exchange ratio (R) possible if the inspired gas contains sufficient nitrogen and net nitrogen exchange is nil. It has been demonstrated that during excercise the onset of anaerobiosis can be identified by a change in R (4). Therefore, monitoring R in critically ill patients may be useful for early recognition of changes in acid-base status.

Methods

We have modified a commercially-available respiratory monitoring system (McGaw Respiratory Therapy, San Marcos, CA) to automati-

27

cally estimate VD/VT and calculate R in mechanically ventilated patients.   In this system, gas analysis is performed by a mass spectrometer capable of measuring oxygen, carbon dioxide, and nitrogen. An electrically-operated valve manifold controls gas input to the mass spectrometer. A gas sampling line, continuously connected to the airway, runs to each bedside. An electronic interface contains analog-to-digital converters for the gas concentration signals, logic circuits for controlling the valves, and a clock. A micro-computer controls operation of the entire system. It was supplied by the manufacturer with programs for calibrating the mass spectrometer with air and a precision gas mixture, analyzing airway gas and calculating, displaying, and recording a number of derived variables, as well as performing a number of offline calculations.

We added a second gas sampling line for each bed, to be continuously connected to a mixing chamber in the expiratory circuit of the ventilator. Advantage was taken of the availability of unused connections on the valve manifold for additional beds to connect these lines.

Figure 1.   Gas Mixing Chamber.

The mixing chamber (Fig. 1) is made of Plexiglas(Reg) 6.3 mm thick. It is 23.6 cm by 16.2 cm by 11.3 cm and has a volume of approximately 3 L. The inlet and outlet fittings fit standard respirator tubing. The outlet is at the bottom of the chamber to permit condenstate to drain. The gas sampling tube is inserted into a fitting near the outlet. The attached clamp permits it to be mounted conveniently on a ventilator.

The program for airway gas analysis supplied with the system is used without substantive modification. The composition is recorded at 0.1 sec intervals for 15 sec. Expiration is identified by examination of the carbon dioxide waveform, the principal criteria being the occurence of a carbon dioxide concentration of 1.5 percent or more and a subsequent value of less than one-half the maximum. The maximum carbon dioxide concentration during the expiration is taken as the end-tidal concentration. The mean of these values for all expirations during the 15 sec interval is converted to partial pressure and recorded as PETCO2. Immediately after the airway gas analysis is completed, gas from the mixing box is routed to the mass spectrometer and mixed expired carbon dioxide tension (PECO2) and oxygen concentration (FEO2) are recorded.

VD/VT is estimated by using PETCO2 as alveolar carbon dioxide tension in the Bohr equation:

$$V_D/V_{T_{est}} = \frac{P_{ET_{CO_2}} - P_{E_{CO_2}}}{P_{ET_{CO_2}} - P_{I_{CO_2}}}$$

R is calculated from this expression (5):

$$R = \frac{F_{E_{CO_2}} - F_{I_{CO_2}} \cdot \dfrac{1 - F_{E_{O_2}} - F_{E_{CO_2}}}{1 - F_{I_{O_2}} - F_{I_{CO_2}}}}{F_{I_{O_2}} \cdot \dfrac{1 - F_{E_{O_2}} - F_{E_{CO_2}}}{1 - F_{I_{O_2}} - F_{I_{CO_2}}} - F_{E_{O_2}}}$$

The nitrogen concentrations are expressed implicitly because it is convenient in our system. In our application R is calculated and recorded each time gas analysis is undertaken provided the inspired gas contains nitrogen. Although the inspired carbon dioxide concentration is ordinarily negligible, we included it in our calculations to make them as generally applicable as possible.

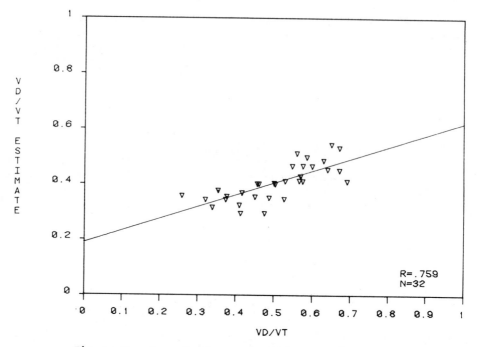

Figure 2. Correlation of VD/VT estimated from end-tidal carbon dioxide tension with VD/VT calculated using arterial carbon dioxide tension. The best-fit linear regression line is shown.

In addition to making the on-line calculations described above possible, routine analysis of mixed expired gas facilitates certain off-line calculations. Whenever we measure arterial blood gases, we enter the results via keyboard, and VD/VT is calculated using the PaCO2. If minute ventilation ($\dot{V}E$) is measured and entered vi˜ keyboard, oxygen consumption is calculated by this relationship:

$$\dot{V}_{O_2} = \dot{V}_E \cdot \frac{F_{E_{O_2}} - F_{I_{O_2}}}{F_{I_{O_2}} \cdot (1-R)-1}$$

This unusual formula is used because, in addition to minute venti- lation, it requires only data that we record for other purposes. Carbon dioxide production ($\dot{V}CO2$) can be calculated thus:

$$\dot{V}_{CO_2} = \frac{V_{O_2}}{R}$$

RESULTS

We determined the completeness of gas mixing in the mixing chamber by analyzing expired gas from mechanically ventilated patients and comparing the result with the result of analysis of a simultaneously collected bag. For 19 such determinations the correlation coefficient was .99.

We have made a limited number of observations comparing the automatically estimated VD/VT with VD/VT calculated using a simultaneously obtained measurement of PaCO2. The results are shown in Figure 2. As expected, estimated VD/VT is considerably lower than VD/VT calculated using PaCO2, but there is a reasonably good correlation, the correlation coefficient being 0.76. The slope of the regression line is 0.43.

DISCUSSION

The degree to which the estimated VD/VT can detect variation in VD/VT in an individual patient is still under evaluation. To the extent that added physiologic dead space behaves in a strictly parallel manner, it will not be detected by our estimate. However, most pathologic conditions causing increased VD/VT produce ventilation-perfusion abnormalities in which the best-ventilated areas empty first. In this case, PECO2 will be reduced more than PETCO2, allowing the change to be detected by our method. For use in monitoring, it is only necessary that the estimate be sufficiently sensitive as an indicator of changes in VD/VT. Variation in the relationship between the two variables among patients is of little consequence. Additional serial observations in patients and animals with ongoing pathophysiologic change are needed to determine whether or not the estimate of VD/VT is sufficiently sensitive as an indicator of change to serve a useful monitoring function. When the ease with which it can be measured is taken into account, a relatively low sensitivity would be sufficient to make it worth while for continuous monitoring. The variety of physiologic changes capable of increasing VD/VT limit its diagnostic specificity. The potential value of monitoring estimated VD/VT automatically probably lies in its ability to warn of cardiopulmonary deterioration, not in diagnosing the cause.

To the extent that the inspired gas mixture contains sufficient nitrogen to permit accurate calculation and nitrogen is not being exchanged, automatic calculation of R presents no difficulty. Calculated R values are not affected by contamination of the expirate by gas from the apparatus dead space, and the required measurements and calculations are easily done. The value of R in monitoring critically ill patients is undetermined. It would seem that a parameter which can detect developing metabolic acidosis during exercise could do so in sick patients as well. Our experience is too limited to determine if this is so, and, to our knowledge, information sufficient to provide an answer has not been reported by others.

ACKNOWLEDGEMENTS

The authors gratefully acknowledge the invaluable engineering contributions of Messrs. Calvin Jackson and William Johnson and thank Ms. Cindy Burris and Mrs. Melinda Smith for typing the manuscript.

REFERENCES

1. LAMY, M., FALLAT, R.J., KOENIGER, E., ET AL.: Pathologic Features and Mechanisms of Hypoxemia in Adult Respiratory Distress Syndrome. Am Rev Resp Dis 114:267, 1976.

2. GERTZ, I., HEDENSTIERNA, G., LOFSTROM, B.: Studies on Pulmonary Function in Patients During Respirator Treatment. Diagnostic and Prognostic Evaluations. Acta Anaesth Scand 20:343, 1976.

3. BARROCAS, M., NUCHPRAYOON, C.V., CLAUDIO, M., ET AL.: Gas Exchange Abnormalities in Diffuse Lung Disease. Am Rev Resp Dis 104:72, 1971.

4. WASSERMAN, K., WHIPP, B.J., KOYAL, S.N., ET AL.: Anaerobic Threshold and Respiratory Gas Exchange During Exercise. J Appl Physiol 35:236, 1973.

5. OTIS, A.B.: Quantitative Relationships in Steady-State Gas Exchange. Eds. W. O. Fenn and H. Rahn. Handbook of Physiology, Section 3: Respiration, Vol I. Washington, American Physiological Society, 1964.

A MICROPROCESSOR CONTROLLED RESPIRATORY MONITOR

A.M. Preszler, G.N. Gee,
R.L. Sheldon, J.E. Hodgkin

Loma Linda University Medical Center
Department of Respiratory Care

BACKGROUND

Development of a respiratory care data collection and monitoring system is based upon the philosophy that it is imperative to have accurate quantitative patient data at the bedside in order to insure quality patient care. Presently, computer systems can adequately provide the means to collection and dissemination of information. The basic consideration becomes one of the quality of the sensor information to be utilized. Primary to respiratory care are the measurements of gas composition, flows, pressures and temperature. New technology is rapidly putting these measurements within reach of the clinicians in the critical care facilities and it is now their responsibility to provide adequate input into the processing of the basic measurements.

The Department of Respiratory Care at Loma Linda University Medical Center (LLUMC) has developed a micro-processor based monitor to continuously measure airway flows, pressures and temperatures. Figure 1 presents the basic structure of the monitor. The transducers are placed in the bi-directional flow area of a patient circuit providing analog voltage levels that are converted into digital values for input to the microprocessor. Computations and waveform analysis are accomplished in real time for breath by breath information display at the bedside in the form of alphanumeric data. In addition, buffered analog outputs are provided for the basic values as well as a serial output port for data storage. Bedside audible and visual alarms are provided. Under an alarm condition, the display is set to a flashing format and gives the description of the alarm condition. The monitor is physically housed in a processor module and a light-weight movable display-control module.

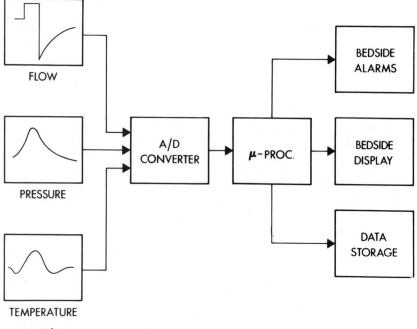

Figure 1: Respiratory monitor functional block diagram.

I. Parameter availabilities
II. Operator interactions
III. Patient interactions
IV. Calibration and maintenance

Figure 2: System considerations for design of a monitor.

System Considerations

The system has been designed following the definition that a "monitor" is a device that continuously measures some patient parameters with no operator or patient cooperation and provides basic alarming mechanisms. There are many pulmonary function tests that can be accomplished when flow, pressure and temperature are known. However, in line with the above definition, they have been excluded from the realm of monitoring. In general, this system analysis to

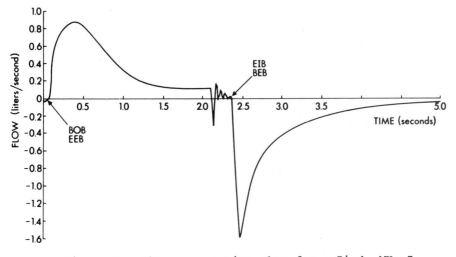

Figure 3:   Flow versus time plot for a Bird   MK   7
Pneumatic Respirator.

provide criteria for the design of a continuous respiratory monitor
must evaluate the considerations outlined in Figure 2.

Parameter Availabilities

Firstly, parameter availabilities are a practical overriding
concern in all cases. For example, it would be clinically advanta-
geous to provide online gas analyzers for continuous monitoring of
airway gas composition. However, the cost, complexity and maintenance of instrumentation to accomplish this purpose when compared
to the monitoring yield, do not justify these measurements. Again,
it must be remembered that in the area of special testing, these
data will be more productive.

The measurement of proximal airway flow is extremely basic to
the support of respiratory care patients and is extremely difficult
to accurately measure under the continuous critical care arrangement.   The productivity of a flow measurement is derived not from
the flow waveform, but from the many secondary parameters that are
calculable once flow is known.   Flow derived variables such as
tidal volumes, minute volumes, respiration rate, compliance and
others when available for patient management on a continuous basis
can be used to provide accurate patient care.   Importantly, flow
becomes the breath delineator that allows accurate calculations of
all breath parameters. Figure 3 shows a flow waveform resulting
from a Bird (pressure controlled) respirator. Plotted are flows in
liters/sec versus time in seconds. Inspiratory flows are positive.

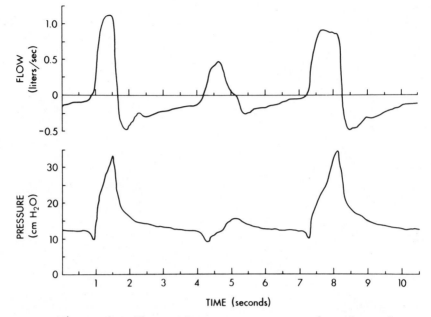

Figure 4:   Flow and pressure versus time plot   for
a Servo 900B volume respirator.

In any system of respiratory monitoring, a most difficult
problem becomes that of identifying the breath in its flow waveform
characteristics.   As seen in Figure 3, beginning of breath (BOB),
end of inspiratory breath (EIB), beginning of expiratory breath
(BEB), and end of expiratory breath (EEB) must be identified for
many diverse flow patterns. Any zero offset of the flow signal can
significantly affect the volume integration.   For example, at a
respiration rate of   12 bpm with a 3 second expiratory time and a
zero offset of 3 lpm, an error of 150 ml/breath is incurred in the
expiratory volume.   At small tidal volumes or longer expiratory
times, this can be a large percentage error, as the divisions
between the breath phases become the most inaccurate. As can be
seen in Figure 3, a ringing of the airflow occurs between the inspiratory and expiratory phases. This is caused in the Bird respirator by the switching of the magnetically controlled
center-body-main-flow-control diaphragm. In this case, the ringing
can cause small errors in volume measurements and large (18%) error
in inspiratory-expiratory times.

Analysis of the pressure waveform is accomplished in conjunction with the flow waveform. The flow waveform provides the delineation of the breath entity. Figure 4 presents a charting of flow and pressure versus time for a patient on a Servo 900B ventilator in the IMV mode. Through correlations with flow, parameters such as PEEP, mean pressure, minimum pressure, etc. are determined. Illustrated in Figure 4 are two types of patient-ventilator breath interactions; a synchronous intermittent mandatory ventilation breath (SIMV) and a patient demand (spontaneous) breath. In the spontaneous breath, the patient receives no direct positive breath support. In the SIMV breath, the patient starts a breath but the ventilator then provides positive pressure. The amount of pressure drop needed to trigger the ventilator and the delay time until the ventilator responds are extremely variable from ventilator to ventilator. The proper measurement of PEEP becomes difficult because of this interaction. As can be seen in Figure 4, before measurable flow occurs, the patient has drawn the pressure below the PEEP value. The monitor analyzes the pressure waveform back in time from the beginning of a breath to determine PEEP and minimum pressure which can provide an indication of the strength of the patient or improper adjustment of the ventilator.

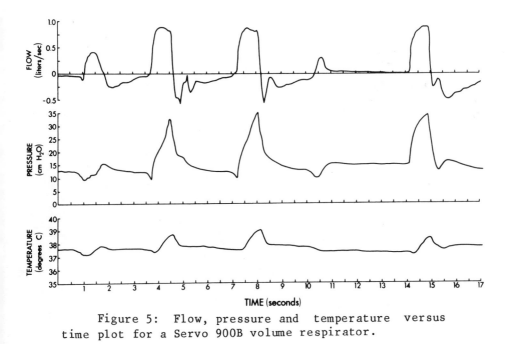

Figure 5: Flow, pressure and temperature versus time plot for a Servo 900B volume respirator.

Temperature waveforms were analyzed to determine the amount of deviation from the average as the air flow through the heated humidifier varies during a patient breath in addition to providing on-line monitoring of patient temperature and average humidified air temperature. Figure 5 is a plot of the flow, pressure, and temperature waveforms for a patient on IMV with a Servo 900B ventilator. Again, flow becomes an important delineator. The patient temperature is derived at that point in time just before the beginning of the inspiratory flow. The heated humidifier excursions are examined during the inspiratory phase. As can be seen from Figure 5, the temperature deviations are about 1.0°C during a breath cycle. The temperature waveform falls upon initial inspiration as the gas that cooled in the tubing, during the previous expiration phase, enters the patient's lungs. An increase in temperature is then recorded as the gas that has been stagnant within the heater system reaches the patient. Subsequently, during expiration, the temperature is seen to decrease to match the patient temperature at end of expiration.

Operator Interactions

The required operator interactions with the monitor have been kept to a minimum to facilitate usage of the system within the constraints of some flexibility of the system, i.e. setting of alarms. The controls consist of two push-switches and one display selector switch. Program control and alarm limit setting are accomplished with the two push switches. Presently, many of the monitoring devices within respiratory care do not provide the operator with information as to their state of calibration. In order to give the user confidence in the system, there are provided verification checks on volume, temperature, and pressure.

Patient Interactions

In line with our definition of a "monitor", no patient cooperation is needed for its operation. The patient interactions occur as problem perturbations to the proper transducer operations. Patient movements, coughing, water and medication condensation, minimum dead space requirements, and so forth are practical problems that must be solved in any monitoring situation.

Calibration and Maintenance

To facilitate long term maintenance, the respiratory monitor incorporates no circuitry unique to our institution. Custom devices, too often, become useless when the original designer leaves for other employment. Therefore, off-the-shelf components have been used in all cases. This provides component interchangability and vendor servicability.

A calibration program is included within the monitor that is available through a switch inside the display housing. Step instructions are displayed that guide the technician in calibrating the voltage levels for flow, pressure, and temperature as well as the ADC zero and gain.

CONCLUSION

The measurement of proximal airway flow is extremely basic to the support of respiratory care patients and is extremely difficult to accurately measure under the continuous, critical care arrangement. However, when available, it can provide the direct measurement of a patient's respiratory parameters. Utilizing the basic measurements of flow, pressure and temperature under the constraints of a continuous monitoring situation, the respiratory monitor display 17 variables as well as differentiating between patient demand and mandatory and synchronous intermittent mandatory breath parameters. It provides six alarm functions visually and audibly.

Without recourse to special testing procedures and patient-therapist interactions, the monitor provides valuable continuous surveillance of the mechanics of respiratory care in the critical care environment.

A MICROCOMPUTER-BASED MULTIPATIENT MONITORING SYSTEM FOR THE RESPIRATORY INTENSIVE CARE UNIT

David Kalinsky, Avram Miller,
Gaby Shapira, Stephen Morris,
M.G. Electronics Ltd.
Kiryat Weizmann,
P. O. Box 102, Rehovot, Israel

Harry Comerchero
Mennen Medical, Inc.
Clarence, New York, U.S.A.

The amount of vital data in the modern Respiratory Intensive Care Unit (RICU) is now so great that conventional monitoring by continuous direct obervation has become impractical. Digital computers offer the capability of automated data concentration and monitoring, as well as the possibility of continuous breath-by-breath analysis of waveforms representing respiratory mechanics and respiratory gas concentrations.

In this context, a microcomputer-based system has been designed and built for monitoring of both hemodynamic and respiratory parameters for a number of patients. The system performs data collection, data presentation and wave analysis functions in the environment of the intensive care unit. The single-processor system can handle an eight-patient ward.

A typical system is based on an LSI-11 processor with disk storage which provides display and user interaction facilities at a central nurse station. Conventional bedside monitors, electronically-controlled ventilators (e.g. Elema Servo Ventilator), and a central mass spectrograph (e.g. Perkin- Elmer MGA) with multiplexed sample lines from each bed, are interfaced to the processor. Data are sampled and analyzed by specially prepared waveform analysis algorithms operating continuously in real time. Respiratory mechanics and gas concentration signal waveforms are sampled at a frequency of 20 Hz each.

The inputs from each of the eight patients in the RICU are as follows:

1. Heart rate value.

2. Arterial blood pressure values (systolic, diastolic, mean).

3. Pulmonary artery pressure values (systolic, diastolic, mean).

4. Central venous pressure value.

5. Patient temperature value.

6. Facemask thermistor waveform or ECG electrodes impedance signal (for derivation of respiratory rate and apnea detection when no other airflow waveform signal is available).

7. Airway pressure waveform signal.

8. Expiratory airflow waveform signal.

9. Airway oxygen concentration waveform signal.

10. Airway CO_2 concentration waveform signal.

From these inputs, the following 16 continuously monitored parameters may be derived and recorded by the system on disk for up to 72 hours.

1. Heart rate (HRT)

2. Arterial blood pressure (ART) (systolic/diastolic, mean).

3. Pulmonary artery pressure (PAP) (systolic/diastolic, mean).

4. Central venous pressure (CVP).

5. Patient temperature (TMP).

6. Respiratory rate (RSP).

7. Ventilatory pressure (VPR) (peak/end-expiratory, end-inspiratory pause).

8. Tidal volume (TVL).

9. Minute volume (MINVOL).

10. Respiratory systm compliance (CMP).

11. Respiratory system airways resistance (RES).

12. Inspired oxygen concentration (FIO2).

13. End-expired CO 2 concentration (FECO2).

14. Oxygen consumption per minute (VO2).

15. Carbon dioxide production per minute (VCO2).

16. Respiratory quotient (including N2 correction).

The parameter values reported are typically filtered averages over several breaths or heartbeats, and are recorded every two minutes for later display as synoptic graphs on video screens. Hard copies may be made of all displays.

22-JAN-79		* WARD STATUS *					22:15:38
BED	HRT	ART	PAP	CVP	TMP	RSP	TVL
1	54	97/ 77	23/17	11	37.4	13	0.66
2	60	112/ 92	30/24	17	38.9	20	0.71
3	57	105/ 85	28/20	11	37.1	13	0.65
4	---	---/---	--/--	--	----	--	----
5	64	108/ 88	28/21	12	37.6	8	0.60
6	57	91/ 71	24/18	11	37.6	19	0.71
7	---	---/---	--/--	--	----	--	----
8	. . .	95/ 75	28/22	14	37.0	. .	. . . .

Fig. 1 Ward Status display. Parameter values in alarm are color-reversed.

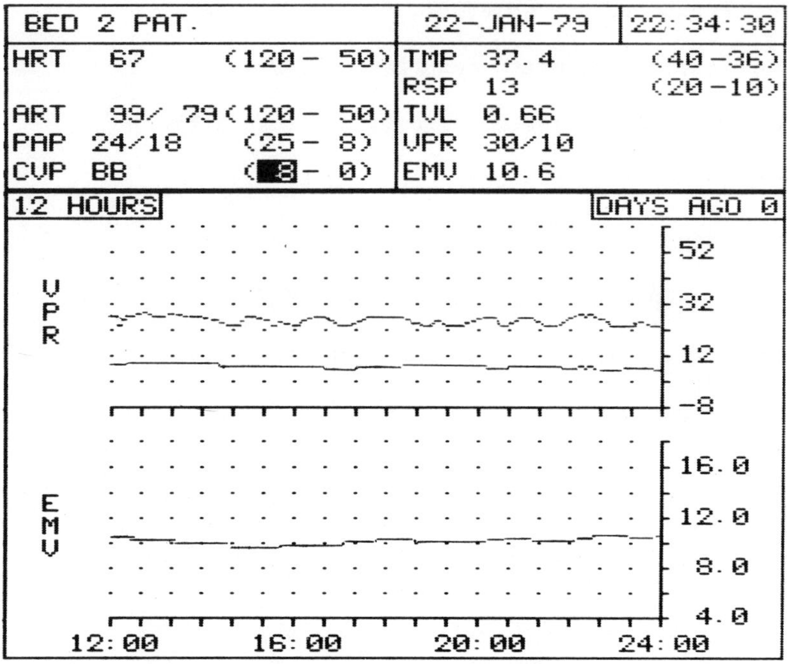

Fig. 2 Interactive display. Trends display of Heart rate and record of MARKS. The vertical lines represent intervention events. The MARKS are labeled at the right.

Current values of the most vital parameters are continuously updated and displayed on the Ward Status screen, shown in Figure 1. This display allows the nurse to survey in concentrated tabular format, the latest values and alarm status (color inversion) for all eight beds. A column is devoted to each of the most vital parameters, and a row for each patient monitored. One such display .s located at the nurse station, and identical slave displays may be scattered throughout the ward.

In addition, the interactive display screen, seen in Figure 2, allows detailed examination of the situation of any one of the monitored patients. This is the second video at the central nurse station, and is operated in conjunction with a simple special-purpose interactive keyboard where a single stroke allows the user to select a patient for detailed review, and short dialogues are used to request a number of interactive functions and displays. Remote terminals which operate identically to that at the

nurse station can be installed in the doctors office and in other
locations. The upper third of this display is reserved for a
'vital signs area' listing the latest continuously updated values
for most derived parameters. Alarm limits are displayed in this
'vital signs area' and may be modified using the interactive keyboard. Continuous monitoring enables improved evaluation and control of both patient and instrumentation, for instance by immediate
reporting of undesirable values for tidal volume, respiratory frequency and respiratory gas parameters.

The bottom two-thirds of the interactive display screen contains an 'interactive area', which serves as a data presentation
and interactive dialogue screen for several user-requested functions.   This area is frequently used for the display of
high-resolution, long-term trends as seen in Figure 2. Trends may
be shown singly or in pairs for any period between 1 and 24 hours
duration, any time during the past 72 hours. All monitored parameters may be trended, with the record of administered interventions,
as seen in Figure 3.

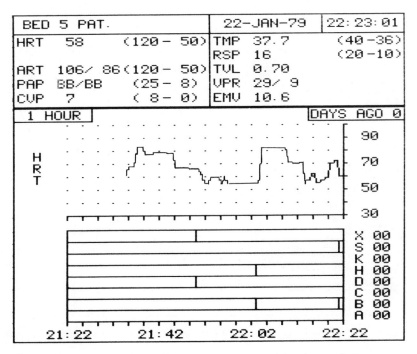

Fig. 3 Interactive display. Trends display of Heartrate and
record of MARKS. The vertical lines represent intervention events.
The MARKS are labeled at the right.

The recording of administered interventions is also carried
out via a user dialogue at the interactive keyboard and display.
This "MARKS" function can be used to record the use of a variety of
drugs and procedures, or even changes in the manner of patient ventilation. The graphic comparison of the 'MARKS' history grid with
the various parameter trends is a powerful tool for evaluating efficacy of treatment.

The actual respiratory waveform analysis algorithms use simple
pattern recognition techniques to detect the medically significant
characteristics of the waveform curves. Respiratory mechanics monitoring is based on the analysis of Airway Pressure and Expiratory
Airflow signals, and the combination of both is required to derive
all the clinically useful information.

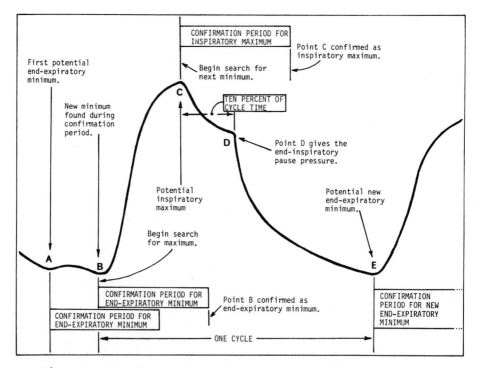

Fig. 4   Algorithm of the Airway Pressure analysis program.

Figure 4 illustrates the technique of Airway Pressure waveform analysis. The major work of the algorithm is the separation of the waveform into "cycles" which correspond to individual breaths. A local maximum and minimum are found, using 'confirmation period' delays to verify that the extrema are the desired ones and not momentary artifacts. The region between two consecutive minima is then called a pressure "cycle", and within each cycle the peak pressure, pause pressure (if there is and end-expiratory pressure) are extracted for that breath. A typical example of the analysis is shown in Figure 5. The upper waveform is the Airway Pressure, annotated with peak position ('P'), pause ('F'), and end-expiration ('V'). The extracted numerical values are seen at the top.

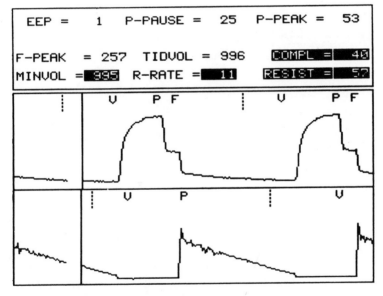

Fig. 5 Example of the analysis of respiratory mechanics. The upper waveform is of Airway Pressure, and the lower of Expiratory Flow. The characters above the waves are computer-produced notations of features (see text). Computed values are seen at the top.

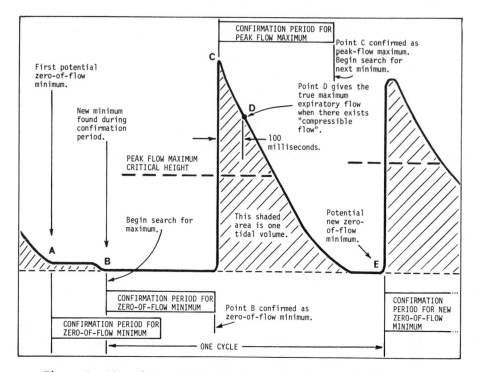

Fig. 6 Algorithm of the Expiratory Flow analysis program.

The lower waveform of Figure 5 is the Expiratory flow. It is analyzed by a similar technique, as seen in Figure 6. However, special care has been taken to eliminate the influence of "Compressible flow" in the tubing between the patient and the transducing device, which may cause an exaggerated maximum. (1) In practice, when measuring flow at the ventilator the "Compressible flow" has a very short-lived effect, and a good approximation for true peak flow can be made by measuring the flow about 100 milliseconds after the detected peak. If flow is measured at the patient, this delay is not implemented. For each 'flow cycle', the true peak flow and instantaneous respiratory rate (from peak-to-peak) are found. In addition, the peak area is calculated (tidal volume) and this area is added to the area of the other peaks over a minute to establish the minute volume.

Values that are unreasonable or that fall outside limits set by the user are reported as special conditions. If both the Airway

Pressure and Expiratory Flow analyses are operating without error conditions, then the Respiratory system compliance and Expiratory airways resistance are calculated on a breath-by-breath basis in real time. If the end-inspiratory pause exists, then the <u>Static</u> compliance is that one which is calculated. Otherwise only dynamic approximations are calculated.

The monitoring facilities described above are presently in routine operation in a number of medical centers. The respiratory gas analysis system is in the final stages of clinical implementation. Again, pattern recognition algorithms analyze breath-by-breath waveforms in real-time; in this case those waveforms represent oxygen and carbon dioxide gas concentrations. A typical example of these analyses is shown in Figure 7.

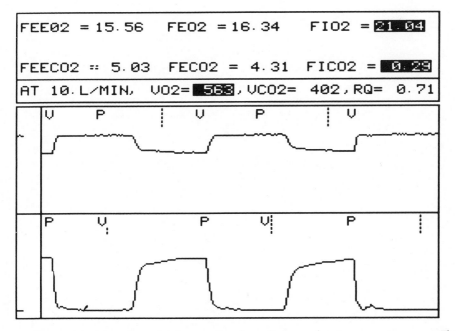

Fig. 7 Example of the analysis of respiratory gases. The upper waveform is of oxygen concentration, and the lower of expiratory flow. The format of this example is as for Figure 5.

In summary, this system enables medical staff to closely monitor the hemodynamic and respiratory conditions of a number of RICU patients. Data may be recorded for inclusion in the patient's files, giving a reliable picture of the patient's progress. Respiratory mechanics and gas waveforms are analyzed and values reported continuously on a breath-by-breath basis in real time.

REFERENCES

1. B. JONSON, L. NORDSTROM, S.G. OLSSON, D. AKERBACK,

        Bull. Physiopathologie Respiratoire 11, 1975.

INSTRUMENTATION AND AUTOMATION OF AN ICU RESPIRATORY TESTING SYSTEM+

Clifford M. Janson, John E. Brimm,
Richard M. Peters

Department of Surgery/Cardiothoracic
University of California Medical Center
San Diego, California

+Supported by USPS Grants GM 17284 and HL 13172, and by Office
of Naval Research Contract N00014-76-C-0282.

ABSTRACT

Since 1971, a commercially available ICU minicomputer system
has been used as the basis for developing an intermittent respiratory testing system. Airway flow and pressure signals and instantaneous gas concentrations, measured by a mobile mass spectrometer,
are used to perform a variety of respiratory function tests at the
ICU bedside. The determinations include respiratory work and mechanics, O_2 consumption and CO_2 production, intrapulmonary shunt
fraction, and spirometry. A telephone communications facility ex-
tends use of the computer to remote hospitals. Data analysis algorithms support the use of intermittent mandatory ventilation
(IMV) and continuous positive airway pressure (CPAP).

BACKGROUND

Pulmonary and cardiac testing is complex and requires sophisticated equipment along with well-trained technicians and
time-consuming calculations. For these reasons, testing has been
traditionally restricted to patients who can go to a laboratory.
In the intensive care unit, where testing can produce information
important to life-and-death decisions, such testing has not been
widely available. With the introduction of two technological inno-
vations, the mass spectrometer and the computer or microprocessor,
ICU testing is not only possible, but cost effective. While analog

51

circuits and hand calculations have been used in the past to generate results, the computer can perform complex calculations for various tests and can easily evolve with new medical and electronic
developments. Computers have been used for respiratory analysis in
the ICU since 1968. (1-10)

EQUIPMENT

     Our computerized ICU respiratory testing system uses a mobile
mass spectrometer (Perkin-Elmer MGA 1100) with attached
Hewlett-Packard pneumotachograph/amplifier and pressure
transducer/amplifier. Either airway or transpulmonary pressure may
be measured. The mass spectrometer measures continuous oxygen,
carbon dioxide, nitrogen, and helium concentrations. When this
"respiratory cart" is wheeled to the bedside, a spring-loaded valve
holds the vacuum in the mass spectrometer for up to 30 minutes.
After transport, power is restored and the cart is operational
within minutes.   The mobility of the cart eliminates the need for
duplicating expensive sensing equipment at each bedside.   Because
the values that we measure relate to lung function, which changes
slowly over time, we utilize intermittent testing, as opposed to
continuous monitoring.

     Currently, testing is performed in six ICU's on four different
floors of the University of California Medical Center, San Diego.
After the cart is wheeled to the bedside, it is plugged into a re-
ceptacle wired to a Hewlett-Packard 5600A ICU data management minicomputer system. We have also developed a "remote" communications
system using telephone links for other hospitals which mirrors the
hard-wired system at our hospital. A microprocessor on a remote
cart is used to control the cart and digitize sampled data; a second microprocessor is used as a front-end data concentrator to re-
lieve loading on our minicomputer. Figure 1 shows a block diagram
of system communications paths. In order to identify carts and patients and to initiate sampling, the remote cart is equipped with a
light-emitting diode matrix and thumbwheel switches under the control of its microprocessor. A special purpose minicomputer driver
written in assembly language controls communication and sampling by
the cart. Thus, one minicomputer installation can support carts at
multiple hospitals. The Trauma Center at the San Diego Naval Regional Medical Center has accessed our minicomputer daily for over
four years. Graphics CRT terminals are used for reporting data in
both the remote and hard-wired communications systems.

CALIBRATION

     On-line calibration of sensing equipment is performed weekly;
several known signals covering the physiological range are sampled

and statistically compared by the computer for linearity and repeatability. Thus, the entire system - transducer, A/D converter, and sampling software - are checked simultaneously. The specific known inputs are a syringe volume delivered at various flow rates, a range of air pressures measured with a water manometer, and samples from precision gas mixture tanks. The advantages of using a syringe instead of a conventional flowrater are that it is faster to set up, is more accurate, and tests volume as well as flow accuracy. The phase lag between flow and gas concentration response is calibrated dynamically as well, using CO_2 readings for breaths with sharp inspirations.

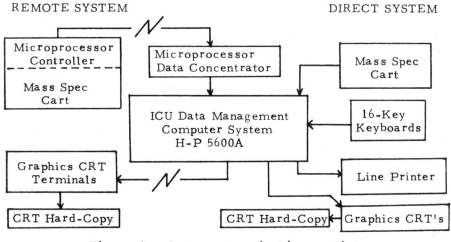

Figure 1.   System communications paths.

TESTS PERFORMED

Our respiratory test battery includes two classes of tests: "passive" tests which do not alter patient breathing patterns, and "active" tests requiring conscious patient cooperation. All of the tests can be performed at the ICU bedside, and are custom designed to allow meaningful use whether respiration is under control of a ventilator or not. Table 1 shows a comprehensive list of tests and the most significant results derived from each one. The clinical significance of some of these results is described in a companion paper. (11).

TABLE 1
RESPIRATORY TESTS PERFORMED

TEST NAME	SIGNALS REQUIRED	DERIVED VALUES
STATIC (STEADY STATE)		
Respiratory mechanics	Flow, pressure (airway) or	Resistance, compliance, Ventilator minute ventilation, Spontaneous minute ventilation, Work of breathing
End-tidal gases	O2, CO2, blood gases	End-tidal O2 and CO2, shunt fraction, dead-space-to-tidal-volume ratio, mean expired O2 and CO2
Gas exchange	Flow, O2, CO2	O2 consumption, CO2 production, respiratory quotient, fick cardiac output
Functional residual capacity	He, CO2	Functional residual capacity
Thermal dilution cardiac output *	Swan Ganz thermal dilution catheter	Cardiac output, stroke work indices, available oxygen
Lung water by double dilution	Densitometer for dye green, thermal dilution catheter	Lung water volume
DYNAMIC (PATIENT COOP.)		
Spirometry	Flow	Forced vital capacity, mean mid-expiratory flow, flow-volume loop graph
Continuous respiratory quotient	Flow, O2, CO2, N2	Continuous respiratory quotient graph
Nitrogen washout	Flow, O2, N2	Closing volume, alveolar plateau slope, residual volume

*Cardiac output is not actually a respiratory test but is included because of its important relationship to lung function.

SPECIAL APPARATUS

Some tests would have been nearly impossible without the design of special apparatus for use with the ventilator. One example
is our functional residual capacity (FRC) measurement box which
utilizes the closed circuit rebreathing technique. (12) Initially,
a 1.5 liter bag is filled with an accurately measured volume of 15
percent helium-oxygen gas mixture, and then switched with a special
purpose valve so the patient is ventilated with this gas mixture.
While in this "rebreathing" mode, the ventilator volume is pumped
around the outside of the bag, which is fully enclosed by a
tight-fitting box. For preoperative and postoperative tests when a
ventilator is not involved, the ventilator inlet port is simply
left open to room air.

During the six or seven breath equilibration period, CO_2 levels are monitored so that they do not exceed 45 mmHg. The test is
considered complete when the end-tidal He concentration is within
0.15 percent of the inspired He during the previous breath.
Initial and final bag volumes and He concentrations are measured
and used to calculate FRC. Repeatability of the test in normal
subjects is within 5 percent.

SIGNAL PROCESSING

Analyses of transmitted data are performed in the 5600A computer by FORTRAN programs. A sampling rate of 25 points per second
per channel of data was proven sufficient for accurately representing respiratory signals. Repetitive sampling of taped data at different rates showed that calculated results did not differ more
than $\pm$ 1 percent.

The first step in processing the raw signals is appropriate
digital filtering. Most programs use a simple 4-point moving average filter. The respiratory mechanics program uses a specially optimized low-pass Butterworth filter in order to eliminate the effects of noise upon theoretical model fitting, particularly with
IMV.

The second step involves rule-based breath boundary picking.
The rules are founded upon two basic criteria for identifying the
inspiratory or expiratory portion of a breath: flow must be of the
proper sign, and a minimum accumulated volume (usually 50 cc's)
must be satisfied. These criteria are sufficient to solve the two
major problems: double inspiration when the ventilator is triggered in IMV, and ventilator oscillation at the end of a breath.

The third step involves correcting breath volumes for the effects of viscosity, temperature, and water vapor. Flow rates measured using pneumotachographs must be adjusted for the viscosity ef-

fects of different gas compositions. Inspired and expired volumes are measured at different temperatures and must be adjusted for comparison. Since the mass spectrometer cannot measure the fraction of water vapor, measured gas concentrations must be corrected depending on whether the gas is wet or dry. In some instances where precise determination of gas volumes is needed, such as in measuring oxygen consumption, inspired and expired nitrogen volumes are used to correct for airway leaks.

The fourth step is classification of breaths into target groups for the test. In the case of pulmonary function tests, the target is simply the largest breath. In the case of single-breath nitrogen washout, the target breath is the one with the first high inhaled oxygen concentration. For the respiratory mechanics program using IMV, a discriminant function is used to distinguish ventilator from spontaneous breaths:

$$F(breath) = maximum \quad (0 \, , \, PV2)$$

where   P = highest inspiratory pressure relative to the start
             of the breath

         V = inspiratory volume.

At the end of its 2-minute data sampling period, the values of F are separated into two groups by examination of their clustering. If the clustering is not significant (i.e., the algorithm failed), the technician is given an opportunity to override the classifications.

The last step involves calculations based on integration of flow to yield volume, flow times concentration to yield gas volumes, and use of techniques such as linear regression and Fourier analysis to fit theoretical curves to the data. Final results and comparison indices are then calculated.

AUTOMATIC SEARCHING

In many instances, results from several different tests and/or hand entries are combined. Programs search within meaningful time windows for all data relating to a particular calculation. For example, shunt fraction may be approximated using arterial blood gas data and hand-entered FIO_2, but venous gases and/or the end-tidal gases (from the mass spectrometer) are used if available within 30 minutes. This type of searching is data driven; that is, the search is initiated whenever any of the relevant data become available or are changed. In this manner, the computer always assures that the most complete information is available.

VERIFICATION

We have found that the experience of skilled respiratory cart technicians is essential to develop confidence in the results calculated by the computer. While algorithms and programs are fully tested before they are used in the ICU, one can never account for all of the variations in human respiratory characteristics, especially in those patients with respiratory diseases. Technicians can not only determine the causes of unexpected results, but provide invaluable insight into test improvements. In order to pinpoint the nature of questionable results, all other possibilities are exhausted before the computer program is re-analyzed. Equipment can be recalibrated to verify its linearity and to check for drift. An oscilloscope on the cart can be used to check the raw signals for peculiarities (such as a reversed pneumotachograph). Finally, the patient record can be examined to determine if unusual results are warranted by the patient's condition.

STORAGE

Our technician-verified final results are stored in the 5600A computer data base, just seconds after the performance of each test. In this manner, we can execute a simple but exhaustive battery of tests in the ICU within just a few minutes' time. The raw data itself may be stored as well, but this facility is used only for program debugging, when the original data are reprocessed to check that a bug has really been corrected. Periodically, the final results from patients who have been "discharged" from the computer are archived onto magnetic tape. These results may then be statistically analyzed using programs which run on larger computer systems.

CONCLUSION

Since our respiratory test package conforms to the specifications of a commercially available medical computing system, its applicability is widespread. The Hewlett-Packard 5600A computer also serves the need of generalized system tasks such as patient admission, file manipulation, and manual information entry such as for blood gas values. Since the only additional hardware necessary to perform the respiratory tests is a facsimile of our cart, use with an existing 5600A computer is feasible. Backed by eight years of use and development, our intermittent respiratory testing system provides a valuable information base for objective clinical decision making in the ICU.

REFERENCES

1. OSBORN, J.J., J.O. BEAUMONT, J.C.A. RAISON: Measurement and monitoring of acutely ill patients by digital computer. Surgery 64: 1057-1070, 1968.

2. OSBORN, J.J., J.O. BEAUMONT, J.C.A. RAISON: Computation for quantitative on-line measurements in an intensive care ward. In Computers in Biomedical Research, edited by Stacy and Waxman, New York: Academic Press, 1969.

3. OSBORN, J.J.: Monitoring respiratory function. Critical Care Medicine 2: 217-220, 1974.

4. DAMMANN, J.F.: Assessment of continous monitoring in the critically ill patient. Dis of Chest 55: 240-244, 1969.

5. LEWIS, F.J.: Monitoring of patients in intensive care units. Surg Clin North America 51: 2815-2823, 1971.

6. HILBERMAN, M., P.J. PATITUCCI and R.M. PETERS: On-line assessment of cardiac and pulmonary pathophysiology in the acutely ill. J. Assoc Adv Med Instru 6: 65-69, 1972.

7. DEMEESTER, J. et al: A method to estimate pulmonary gas volume, distribution of ventilation and pulmonary capillary blood flow in intensive care. Computers in Cardiology, Long Beach: IEEE Press, 205-209, 1975.

8. WILLIAMS, M.J., J.W. RUBIN and R.G. ELLISON: Calculator-controlled respiratory ICU system. Computers in Cardiology, IEEE Press, pp.231-235, 1976.

9. TURNEY, S.F., B.S. BLUMENFELD, S. WOLF and R. DERMAN: Respiratory monitoring. Ann Thorac Surg 16: 184-192, 1973.

10. BRIMM, J.E., C.M. JANSON, M.R. O'HARA, P.J. PATITUCCI and R.M. PETERS: An automated testing system for cardiopulmonary function. Computers in Cardiology, Long Beach: IEEE Press, 1978. Pp. 335-338.

11. PETERS, R.M., J.E. BRIMM and C.M. JANSON: Clinical basis and use of an automated ICU testing system. Proc. First Annual International Symposium in Critical Care and Pulmonary Medicine, Norwalk, Connecticut May 24-26, 1979.

12. HELDT, G.P. and R.M. PETERS: A simplified method to determine functional residual capacity during mechanical ventilation. Chest 74: 492-496, 1978.

MONITORING OF THE CYCLIC MODULATION OF CARDIAC OUTPUT DURING

ARTIFICAL VENTILATION

Jos. R.C. Jansen, Jan M. Bogaard,
Eric von Reth*, Jan J. Schreuder,
Adrian Versprille

Pathophysiological laboratory,
Department of Pulmonary Diseases
Erasmus University
Rotterdam, The Netherlands

*Physics Department
Eindhoven University of Technology
Eindhoven, The Netherlands

The cardiac output (CO) is one of the main factors which determine oxygen delivery to the tissues. For its estimation several methods exist, which all have disadvantages under different experimental and clinical conditions. The usage of the thermodilution for the estimation of the CO was introduced by Fegler (1). He and others have demonstrated the feasibility of this method in animals (2, 3, 4, 5) and men (6, 7, 8). The measurements are easy to perform and can be repeated almost without limitations at very short time intervals. In combination with a computer the cardiac output can be presented within 30 sec. after injection of the indicator. However, the application of a dilution technique is limited due to a few important conditions (9, 10, 11, 12).

1. Loss of indicator between injection and measuring place is not allowed to occur, which might be of importance, when using temperature as an indicator.

2. A constant pulsatile blood flow during the period of a measurement has to be present.

3. The indicator has to be completely mixed with the blood before the first branch and measuring place.

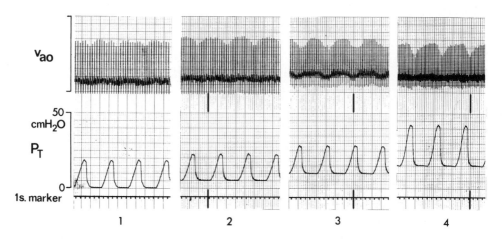

Fig. 1:   Effects of PEEP on stream velocity in the
aorta (vao) and tracheal pressure (PT) at a PEEP level
of 0, 0.5, 1.0 and 1.5 kPa respectively. vao is presented in arbitrary units, which are equal for all 4
measurements.

Under circumstances of artificial ventilation the cardiac output is modulated during each cycle (13), which is not in agreement
with the second condition.

In order to estimate the errors of the thermodilution measurements   due   to   these   cyclic changes of the cardiac output we performed a series of measurements at the left side of the heart under
increasing levels of positive and expiratory pressure (PEEP). This
PEEP was used in order to change cardiac output as well as the   amount of modulation, which is illustrated in fig. 1 by the stream
velocity in the aorta in relation to increased tracheal pressure by
PEEP.   Under these circumstances the systematic deviations of the
individual measurements with respect to the mean value of the   cardiac output were analyzed. From this analysis the best moment of
injection was defined for the estimation of the   average   value   of
the cardiac output.   In order to evaluate the accuracy of the thermodilution method we compared the mean   values   with   corresponding
values determined by the direct Fick method for oxygen.

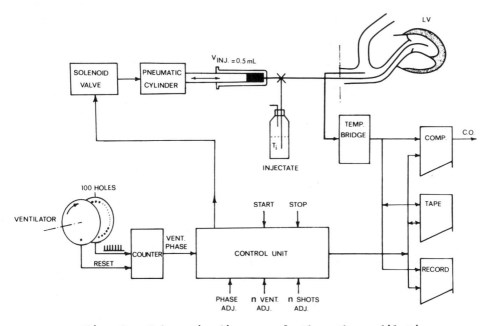

Fig. 2: Schematic diagram of the thermodilution
technique.

Methods

In five pigs several series of 50 measurements were carried
out under steady state conditions at 50 equidistant moments of the
ventilatory cycle. The sequence of the phases was chosen at random. At each level of PEEP such a series of 50 measurements was
performed. In order to control the steady state haemodynamically
throughout such a series not only arterial and central venous blood
pressure were measured, but also CO according to the direct Fick
method for oxygen. Three such measurements of CO were done before,
respectively, after each series. The blood and gas samples were
collected over a period of about three respiratory cycles.

All animals were ventilated with a Starling pump at a rate of
10 per minute. The inflation lasted 44% of the cycle period and
was followed by a spontaneous expiration against a water-seal from
0 up to 1.5 kPa. The thermodilution measurements were performed by

injection of 0.5 ml saline at room temperature into the left ventricle and by detection of the dilution curve in the aortical arch (fig. 2). The saline was injected within 300 msec by means of a pneumatic syringe through a double-walled catheter with three openings at the tip. After each measurement the injector was automatically refilled. Volume reproducibility was checked. The pneumatic cylinder was driven by compressed air and initiated by an electric signal of the control unit.

The moment of injection was depending on two pre-set factors:

1.   the moment in the ventilatory cycle, and

2.   a certain number of ventilations between 2 injections; in our experiments this number was five.

The moment of injection in the ventilatory cycle was derived from two discs on the axis of the Starling ventilator, one with 100 holes and the other with one hole. The disc with 1 hole was used to reset the counter at the start of the inspiration. The disc with 100 holes divides the ventilation cycle in percentages. The control unit controlled the moment of the injection, and signalled it to the computer, the tape recorder and the paper recorder. From this moment the computer started the integration of the temperature time curve, derived from the temperature bridge.

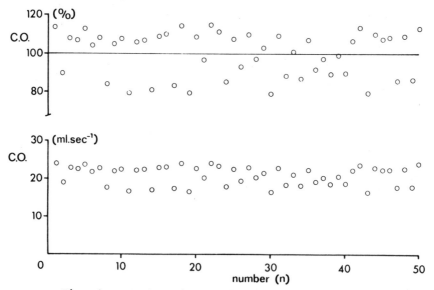

Fig. 3.a.: A series of 50 cardiac output measurements (ordinate) in one animal plotted agianst the serial number of the injection (abscis).

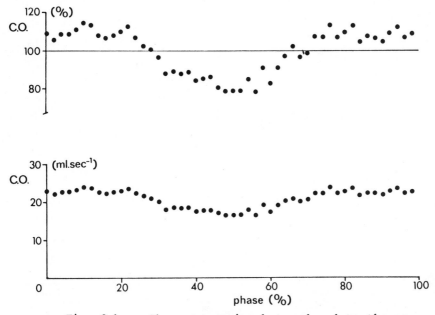

Fig. 3.b.: The same series but ordered to the moment of injection, as a phase of the ventilatory cycle.

The cardiac output was calculated by using the
Steward-Hamilton formula, modified to calculate cardiac output by
thermodilution (8). In this calculation two important corrections
were made; one for the leakage of injectate through the wall of
the injectioncatheter, and the other for the respiratory fluctuations in the base-line. These fluctuations were evaluated before
the calculation of each series of cardiac output measurements.

RESULTS

In fig. 3.a. an example of a series of measurements is presented with all individual values ordered according to the serial
number of injection. There is no trend of cardiac output with
time, i.e. the order of injection, thus a steady state might be
accepted. The mean value was calculated from all 50 measurements
and was taken as the 100% value. Each individual measurement is
expressed as a percentage of this mean value. The maximal difference between two measurements within one series is about 40%.

A cyclic modulation of the values appears after sorting the
measurements with respect to the moment of injection as a phase of
the ventilatory cycle (fig. 3.b.). On this modulation a random
error is superimposed. In fig. 4 all results of five animals are
averaged. At a PEEP level of 1,5 kPa the results of only 4 animals
were used, because in one case the steady state was not present
under these circumstances.

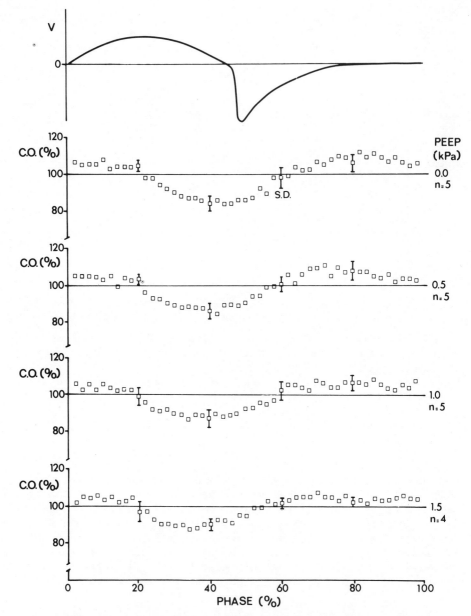

Fig. 4: The cyclic modulation of the CO at different levels of PEEP of 5 pigs. As a reference the airflow is given at the top. The vertical bars at a ventilationphase of 20, 40, 60 and 80% represent the standard deviation of the mean.

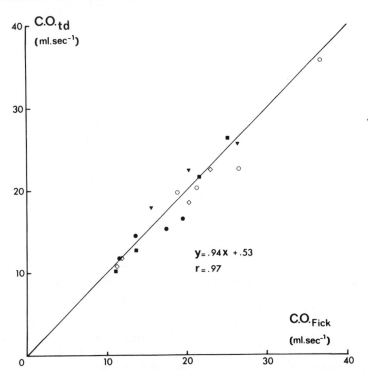

Fig. 5: Comparison of cardiac output measured
with the thermodilution technique (COtd) and the direct
Fick method for oxygen (COFick). The plot includes 19
series of measurements in 5 different pigs. The line
is the identity line.

During application of PEEP, cardiac output decreased with respect to the mean value at a PEEP level of zero.

The relative values of the cardiac output at PEEP levels of
.5, 1.0 and 1.5 were respectively 82% (SD +- 7%), 59% (+- 4%) and
51% (+- 5%). For comparison of the thermodilution method with the
Fick method the mean value of each series was plotted against the
mean values of the two series of three Fick-measurements respectively before and after each thermodilution series (fig. 5). The
line in fig. 5 is the identity line. The equation of the regression line is y = .53 + .94x with a correlation coefficient r = .97.

DISCUSSION

One of the three main sources of errors in the estimation of cardiac output by means of the thermodilution technique, the modulation of blood flow, was tested in the investigations presented in this paper. The other two sources of errors, being poor mixing and loss of indicator, were not systematically studied here. The mixing was accepted to be complete because of the relatively low random error of the measurements as seen from fig. 3.b. According to Warren and Ledingham (3) the loss of indicator within the left ventricle and the ascending aorta might supposed to be very small, which is confirmed by the very good correlation of the mean values of the thermodilution and Fick measurements for each series of observations as presented in fig. 5. It is obvious from the scatter of points in this figure, that no statistical tests are needed in order to indicate that both methods give the same results.

From our results it appeared that measuring cardiac output and neglecting the phase of the ventilatory cycle a difference of about 40% between two measurements might be found. This conclusion only counts for the usage of the indicator dilution technique at the left side. However, Nordstrom (14) and Morgan (13) showed also very large modulations of flow at the right side of the heart under circumstances of artificial ventilation. Therefore, we suppose that the same conclusion counts for using the technique at the right side of the heart. If the measurements are plotted against the moment of injection as a phase of the ventilatory cycle, the large range of 40% changes into a systematic modulation with a random error superimposed on it. At the right side of the heart we might expect a similar phenomenon, however with a difference in the amount and a shift the phase of the modulation. For an estimation of the mean value of cardiac output the injections at about 60% of the ventilatory cycle appeared to be the best moment. This is true for our experiments but this has not to be necessarily valid for measurements under other conditions of artificial ventilation. Nevertheless, we think that at the end of spontaneous expiration on the mean value of cardiac output might be estimated most accurately; at that moment the venous return to the left heart has been restored. The amount of modulation is about the same at all levels of PEEP (fig. 4). Although the modulation of flow, demonstrated as velocity in fig. 1, increases with PEEP, this increase is not measured by means of the thermodilution technique. We must be aware of the fact that every individual thermodilution measurement represents a certain mean of all actual flows during the time of the procedure. So, the flow values in the first part of each dilution curve contribute the most to this mean and the flow values during the tail of the curve progressively less. Therefore, the value of the cardiac output will be some weighed mean of all these flow values, which means that the modulation of the cardiac output during a ventilatory cycle might be expected to be greater than is estimated by the thermodilution technique. Although we expected

the averaging effect on the estimation, because of the weighed influences of all flows during the time course of the dilution curve,
it nevertheless was surprising to us that we did not observe increasing difference in modulation with PEEP. In conclusion, the main
results of our work are:

1. that according to the comparison with the Fick method the
    thermodilution method can give reliable estimates of car-
    diac output, when injecting the indicator into the left
    ventricle and detecting it in the aortic arch,

2. that this technique is useful for frequent monitoring
    under changing circumstances,

3. that under circumstances of artificial ventilation the
    maximal differences between two measurements at any moment
    of the ventilatory cycle might be found to high for an ac-
    curate estimation of the mean flow,

4. that the best moment of injection seems to be at the end
    of spontaneous expiration, if the measurement is performed
    at the left side of the heart.

REFERENCES

1. FEGLER, G.: Measurement of cardiac output in anaesthetized an-
    imals by a thermodilution method. Quart. J. Exp.
    Physiol. 39, 153-164, 1954.

2. KHALIL, H.H., T.Q. RICHARDSON and Q.C. GUYTON: Measurement
    of cardiac output by thermal dilution and Direct Fick
    methods in dogs. J. Appl. Physiol. 21, 1131-1135,
    1966.

3. WARREN, D.J. and J.G.G. LEDINGHAM. Cardiac output in the
    conscious rabbit: An analysis of the thermodilution
    technique. J. Appl. Physiol. 36, 246-251, 1974.

4. SILOVE, E.D., T. GANTEZ and B.G. WELLS: Thermodilution meas-
    urement of left and right ventricular outputs.
    Cardiovascular Res. 5, 174-177, 1971.

5. GOODYER, A.V.N., A. HUVOS, W.F. ECKHARDT and R.H. OSTBERG:
    Thermal dilution curves in the intact animal. Circ.
    Res. 7, 432-441, 1959.

6. BRANTHWAITE, M.A. and R.D. BRADLEY: Measurement of cardiac output by thermal dilution in man. J. Appl. Physiol. 24, 434-438, 1968.

7. OLSSON, B.,J. POOL, P. VAN DER MOSTEN, E. VARNAUSKAS and R. WASSEN: Validity and reproducibility of determination of cardiac output by thermodilution in man. Cardiology 55, 136-148, 1970.

8. GANZ, W. and H.J.C. SWAN: Measurement of blood flow by thermodilution. Am. J. Cardiol. 29, 241-246, 1972.

9. VLIERS, A.C.A.P.,K.R. VISSER and W.G. ZIJLSTRA: Analysis of indicator distribution in the determination of cardiac output by thermodilution. Cardiovascular Res. 7, 125-132, 1973.

10. CROPP, J.A. and A.C. BURTON: Theoretical considerations and model- experiments on the validity of indicator dilution methods for measurement of variable flow. Circ. Res. 18, 26-48, 1966.

11. WESSEL, H.U. M.H. PAUL, G.W. JAMES AND A.R. GRAHN: Limitaitons of thermal dilution curves for cardiac output determination. J. Appl. Physiol. 30, 643-652, 1971.

12. SCHEUER-LEESER, M. A. MORQUET, H. REUL AND W. IRNICH: Some aspects to the pulsation error in blood-flow calculatons by indicator- dilution techniques. Med. Bio. Eng. Comput. 15, 118-123, 1977.

13. MORGAN, B.C. W.E MARTIN, T.F. HORNBEIN, E.W. CRAWFORD AND WG GUNTHEROTH: Hemodynamic effects of intermittent positive pressure respiration. Anaesthesiol. 27, 584-590, 1966.

14. NORDSTROM, L: Haemodynamic effects of intermittent positive pressure ventilation with and without an end-inspiratory pause. Acta anaesth. scand., suppl. 47, 1972.

COMPUTER ASSISTANCE IN ASSESSMENT AND MANAGEMENT OF MECHANICAL VENTILATION

Steven A. Conrad', M.D.
Ronald B. George*, M.D.

'The Department of Biometry,
Case Western Reserve University
School of Medicine,
Cleveland, Ohio

*Department of Medicine,
Louisiana State University
Medical Center,
Shreveport, Louisiana

Supported in part by
National Library of Medicine grant 1-T15-LM07001

INTRODUCTION

The decision to institute artificial mechanical ventilation is undertaken when a patient's pulmonary system is unable to maintain normal function unaided. The goal is to achieve near-normal gas exchange and blood gas levels with the ventilator functioning in a manner similar to normal physiologic ventilation. To enable this, the modern ventilator offers great flexibility in ventilation, oxygenation, and mode of operation.

The physiologic principles underlying the management of patient's receiving mechanical ventilation are quantitative and have been for the most part well defined. The application of these principles to ventilator management is hindered by their complexity and by the difficulty in obtaining many indices. As a result, ventilator operation is often performed emperically, and adjustment for an optimal state is accomplished by repeated blood gas measurement. To provide assistance in ventilator management, we are developing a computer program designed to enable rapid attainment and maintenance of a stable condition, a reduction in the number of arterial blood samples drawn, and possibly a reduction in the time that the patient is supported by the ventilator.

OBJECTIVES

We defined four major roles of a computer in this application:

1. assist with collection, storage, retrieval, and display of relevant data.

2. provide an assessment of the patient's status by evaluation of readily available blood gas and ventilation data.

3. assist in maintaining patients in an optimal state while on the ventilator, and minimize time they spend on it.

4. provide surveillance over the many indicators of the patient's condition over time.

METHODS

The program is an off-line system designed for interactive data entry by persons with minimal computer experience, typically the technician who performs the blood gas determination. Output is in report format suitable for inclusion in the patient's medical record if desired. The content of the report is in accord with the type of input data, but basically consists of sections of measured and derived parameters followed by recommendations. The program is written in FORTRAN IV in order to provide portability between computer systems.

The system performs the following general functions, implemented as individually selectable procedures:

1. data entry, validation, and storage

2. arterial and venous blood gas analysis and interpretation

3. ventilator setup assistance

4. ventilation and weaning assessment and recommendations

5. data retrieval and display

Data Entry

Two general types of information are requested by the system.
1. At the initial encounter, the patient is registered with the
system. Patient characteristics, such as age, weight, and sex as
well as ancillary information such as ventilator type are stored by
the system. 2. Blood gas and ventilation data are entered as they
are collected. All data previously entered are available for any
analysis. Data entry at each encounter is flexible, with the program giving the appropriate prompts. Each entry is compared against physiologic limits for detection of possible errors. Visual
verification is enabled at a second point before each section is
submitted.

Blood Gas Analysis

The blood gas analysis procedure requires arterial pH, PCO2,
PO2, hemoglobin, hematocrit, and body temperature. The measured
values are corrected for temperature and then used to calculate
standard parameters, among which are actual and standard bicarbonate, base excess, O2 content and saturation, and an estimate of
the P50 (1). An assessment of the acid-base status is made according to a modification of the acid-base map of Cohen (2). This procedure uses PCO2 and base excess to define 33 different interpretation areas. An oxygenation assessment is also made, concentrating
on oxygen delivery as well as PO2.

Venous blood gas analysis values are accepted if available,
and contribute to calculations of physiologic shunt and cardiac
output.

Ventilator Setup

The setup procedure enables calculation of several options of
rate, tidal volume, and FiO2 for initial ventilator operation,
based on patient characteristics and blood gas data. The expected
basal ventilation is calculated according to the Radford method

(3), and modified by the patient's physiologic dead space fraction (Vd/Vt) derived from pre-ventilator indices if available. The goal is to place the patient at an initial state of optimal alveolar ventilation and oxygenation with as little uncertanity as possible.

Ventilation Analysis

The ventilation analysis section involves the calculation and reporting of useful indices of ventilation, assessment of the status of the ventilatory state of the patient, and formulation of recommendations in the areas of ventilation, oxygenation, and weaning. The input values are directly obtainable from the ventilator and include respiratory rate, tidal volume, peak and pause pressures, peak flow, external dead space, PEEP, and assist/control mode. Corrections are made in delivered volume for system compliance. Calculation of total, elastic, and non-elastic resistance enables monitoring of acute changes in the airway and in lung compliance. Determination of alveolar PO2 (PAO2) enables the calculation of A-a gradient and a-A ratio which are used in following the oxygen transfer ability of the lungs. The calculated parameters include an estimate of the Vd/Vt by the Levesque method (4) as modified by Atkin (5):

$$\frac{Vd}{Vt} = 1 - \frac{Ve}{Vm} \times \frac{40}{PaCO_2} \times .67$$

where Vm is measured minute ventilation and Ve is the expected minute ventilation. This procedure decreases uncertainty in control of alveolar ventilation.

Expired gas values may be input for Vd/Vt calculation by the Bohr method.

Recommendations for changes in ventilator settings are based upon the assessment of the patient's condition and the current ventilator settings in relation to an optimal physiologic state. We define an optimal state as one in whch the level and mode of administration of minute ventilation are appropriate for the metabolic state of the patient. The desired PaCO2 for a given patient is chosen as that which would be expected if the patient's ventilatory function were normal. If the patient's acid-base state is normal, the desired PCO2 (PdCO2) is chosen as the patient's normal PCO2. In the case of metabolic alkalosis or acidosis, the patient is slightly hypo- or hyperventilated. Rather than aim for a single desired pH value, the intent is to simulate a chronic compensatory state. This gives a pH which varies from normal with the metabolic

disturbance. If the patient has chronic obstructive pulmonary disease, the equation is modified slightly to reflect a normally high
$PCO2$. The basic equations are:

$$PdCO2 = 0.8 \times BE + 40 \qquad BE > +3$$

$$PdCO2 = 1.1 \times BE + 40 \qquad BE < -3$$

where BE represents the base excess. The desired $PCO2$ obtained is
compared to the current $PCO2$ and is used to calculate changes in
rate, tidal volume, or external dead space. One, two, or all three
of these may be used in the recommendation.

The actual recommendation, provided in one of 21 different
formats, is dependent upon several factors: the assist/control
mode, the proximity of the current rate and tidal volume to normal
values, the resistance of the lungs, and the amount of external
dead space. Associated with the recommendations is a brief expla-
natory note as to why the change is required.

The proposed changes in rate are fairly straightforward in
calculation, since within limits, alveolar ventilation and $PCO2$ are
directly proportional to change in rate. If appropiate, external
dead space calculations are based on the method of Suwa and Bendix-
en (6).

A non-trivial relationship is the one between $PCO2$, alveolar
volume, and tidal volume. For accurate changes to be calculated in
tidal volume, consideration must be made of the anatomic and alveolar dead space, the external dead space, and the loss of volume due
to compliance. The tidal volume, as delivered by the ventilator,
is assumed to consist of the following components:

$$Vt = Va + Vdal + Vdan + Vde + Vcl + Vcs$$

The alveolar ventilation (Va) changes proportionally with $PCO2$, the
anatomic dead space (Vdan) and external dead space (Vde) are assumed constant, and the alveolar dead space (Vdal) is assumed (from
work by Lifshay et al (7) to increase linearly with alveolar volume. The volume lost in the lung due to compression is determined
from the pressure generated. Volume loss in the ventilator system
is determined from system compliance and pressure. Using these assumptions, a tidal volume intended to give a desired alveolar ventilation and $PCO2$ may be accurately predicted.

In addition to ventilation, recommendations are made with regard to oxygenation. It has been demonstrated that the
arterial-alveolar $PO2$ ratio remains fairly constant over moderate
ranges of $FiO2$ (8). This forms the basis of the calculations of
the $FiO2$ required to produce a desired $PaO2$.

Weaning recommendations are included in each patient report. The program examines the blood gas values, respiratory drive, indices of ventilation-perfusion relationships, and resting and forced ventilatory measurements. If successful weaning is not likely, the reasons are reported.

To provide some insight into the relationship of the blood gases to ventilation, the program can produce a plot of PaO2 and PaCO2 as a function of minute ventilation, based on the steady-state respiratory models of Milhorn and Brown (9) and the assumptions developed above with regard to tidal volume. This enables the user to see the effect of minute ventilation at values other than the one recommended.

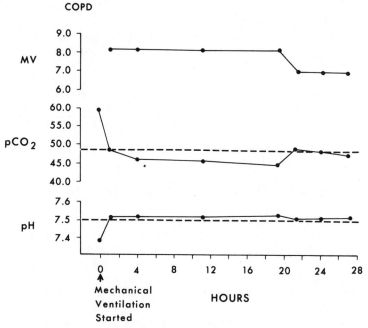

Figure 1. The time course of the PaCO2 and pH of a patient being mechanically ventilated with computer-recommended changes in minute ventilation (MV) at 0 and 19 hours. The desired values are indicated by the dotted lines. The initial PaCO2 and pH values are pre-ventilator measurements.

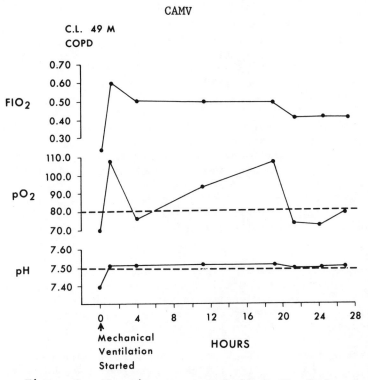

Figure 2.   The time course of the PaO2 of the same
patient   present   in Fig. 1.   The initial values of the
three variables represent pre-ventilator   measurements.
Desired values are indicated by the dotted lines.

In addition to the recommendations, the program provides   surveillance   for significant changes in key variables.   These include
Vd/Vt and a-A ratio for ventilation-perfusion relationships, pH and
base   excess   for metabolic disturbances, and resistance for airway
problems.

RESULTS

Full evaluation of the accuracy and efficacy of the program is
still   being   performed, since we have so far used the program with
only a small number of patients.   The results of one of these patients   are   given in Figs.   1 and 2.   The ability to bring the PaCO2
and PaO2 to the desired level after a recommended change in   minute
ventilation or FiO2 with small tolerance is demonstrated.   The greatest error occurred in the prediction of the   FiO2   at   ventilator
setup;   this is due to an inabilty to predict the change in ventilation-perfusion   relationships   between   natural   and
positive-pressure ventilation.

DISCUSSION

Accurate control of ventilation in patients requiring continuous support has been indentified as a problem. Approaches to this problem have included trial-and-error management, the use of graphs and nomograms, and the application of known relationships by the use of equations.

The trial-and-error method, based on the inverse relationship between PCO2 and minute ventilation, and the relationship between FiO2 and PaO2, is easily applied, but requires repeated blood gas determinations and a longer time to reach an optimal state.

The use of nomograms, such as those advocated by Radford et al (10) and Selecky et al (11), are a bit more accurate, but still do not account for many of the variables important in ventilator management.

The use of a computer system to make complex calculations and decisions increases the utility of readily available parameters of ventilation in managing ventilator patients, while freeing the physician from time-consuming calculations. In addition, it provides a comprehensive information management facility.

Previous investigations into computer-aided management of mechanical ventilation (12, 13) have been reported, but either have not been comprehensive or have not dealt fully with the question of clinical efficacy. Our initial results with a limited number of patients are promising, and we hope to determine the long-term effect on the clinical course of the patient as well as to evaluate the accuracy of the system.

The authors wish to acknowledge the assistance of Mr. William H. Kincaid.

REFERENCES

1. THOMAS LJ: Algorithms for selected blood acid-base and blood gas calculations. J Appl Physiol, 33(1): 154-158, 1972.

2. COHEN ML: A computer program for the interpretation of blood-gas analysis. Comp Biomed Res, 2: 549-557, 1969.

3. RADFORD EP: Ventilation standards for use in artificial respiration. J Appl Physiol, 7(4): 451-460, 1955.

4. LEVESQUE PR AND ROSENBERG H: Rapid bedside estimation of wasted ventilation (Vd/Vt). Anesthesiology, 42(1): 98-100, 1975.

5. ATKIN DH: Estimating wasted ventilation (letter to the editor). Anesthesiology, 43(1): 128-129, 1975.

6. SUWA K AND BENDIXEN HH: Change in PaCO2 with mechanical dead space during artificial ventilation, J Appl Physiol, 24: 556-563, 1968.

7. LIFSHAY A, FAST CW AND GLAZIER JB: Effects of changes in respiratory pattern on physiological dead space. J Appl Physiol, 31(3): 478-483, 1971

8. GILBERT R AND KEIGHLY JF: The arterial/alveolar oxygen tension ratio. An index of gas exchange applicable to varying inspired oxygen concentrations. Amer Review Resp Dis, 109: 142-145, 1974.

9. MILHORN HT AND BROWN DR: Steady-state simulation of the human respiratory system. Comp Biomed Res, 3: 604-619, 1971.

10. RADFORD EP, FERRIS BG AND KRIETE BC: Clinical use of a nomo-
     gram to estimate proper ventilation during artificial
     ventilation. New England Journal of Medicine, 251(22):
     877-884, 1954.

11. SELECKY PA, WASSERMAN K, KLEIN M, AND ZIMENT I: A graphic ap-
     proach to assessing interrelationships among minute ven-
     tilation, arterial carbon dioxide tension, and ratio of
     physiologic dead space to tidal volume in patients on
     respirators. Amer Review Resp Dis, 117: 181-184, 1978.

12. MENN SJ, BARNETT GO, SCHMECHEL D, et al: A computer program
     to assist in the care of acute respiratory failure.
     JAMA, 223(3): 308-312, 1973.

13. GROSSMAN RF, HEW E, AND ABERMAN A: Validation of a computer
     program to manage patients on mechanical ventilators.
     Intens Care Med, 3(3):211, 1977.

CLOSED LOOP THERAPY IN THE CRITICAL CARE UNIT USING MICROPROCESSOR

CONTROLLERS

D.R. Westenskow, R.J. Bowman,
D.B. Raemer, K.B. Ohlson,
W.S. Jordan

Departments of Anesthesiology and
Bioengineering
University of Utah
Salt Lake City, Utah 84132

Automatic closed loop control of therapeutic interventions in the critically ill patient allows therapy to be optimized to rapidly changing patient needs. Three such examples of closed loop control in clinical care are presented: 1) the mechanically ventilated patients where the rate of ventilation is controlled according to metabolic rate, 2) the burn patient where IV fluids are given to maintain urine output and 3) a method of monitoring metabolic rate using a feedback replenishment technique. In each case a microprocessor system provides the necessary feedback control and user interaction. The microprocessor itself is small and inexpensive enough to provide integration of multiple inputs, high level user interaction, and flexibility during instrumentation development.

I.   Control of Ventilation

The mechanically ventilated patient will benefit greatly if his ventilation is controlled by a closed loop feedback system so that ventilation changes as his metabolic needs change.   We have developed a microprocessor system which samples CO_2 production and adjusts the rate of ventilation to keep the patient's arterial CO_2 concentration ($PaCO_2$) constant.

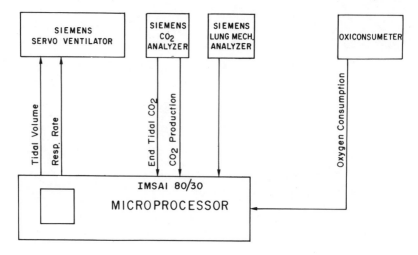

Figure 1.

A block diagram of the system is shown in Figure 1. An IMSAI 80/30 microprocessor system is used for acquiring data and controlling the ventilator. This system samples 16 input parameters with 12 bit accuracy, displays data and trends on a 5 inch CRT and produces the signals for controlling the ventilators rate and tidal volume. The Siemens 900B Servoventilator is used with the 930 CO_2 analyzer and 940 lung mechanics calculator. Signals sampled from the Siemens system include CO_2 production ($\dot{V}CO_2$), ineffective tidal volume (VD), end tidal CO_2 concentration and lung compliance. The oxygen consumption ($\dot{V}O_2$) comes from an oxiconsumeter. (1)

The algorithm used to control ventilation is similar to that proposed by Mitamura et al. (2) The alveolar ventilation ($\dot{V}A$) necessary to maintain a desired alveolar concentration of CO_2 (PACO2) is given by:

$$\dot{V}_A = \left[\frac{P_B - 47}{P_A CO_2} \right] \dot{V}CO_2$$

where PB is the barometric pressure.
Minute ventilation ($\dot{V}E$) can be calculated from:

$$\dot{V}_E = k\,\dot{V}_A + V_D \cdot f$$

where f is the frequency of ventilation and k a constant related to physiologic dead space. Closed loop control is discontinued if the $\dot{V}CO2/\dot{V}O2$ ratio becomes less than 0.7 or greater than 1.1 or if the end tidal CO2 becomes greater than 7.0 or less than 4.0.

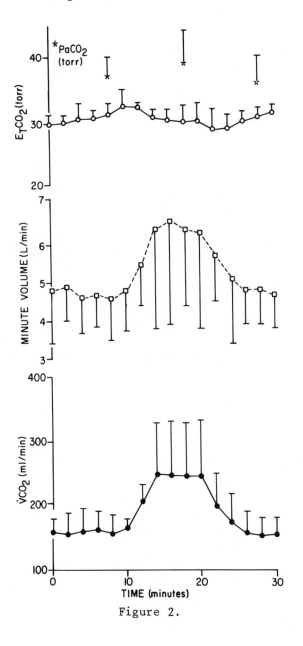

Figure 2.

Results are shown in Figure 2. Three dogs (20-25 kg) were anesthetized with sodium thiopental, maintained with fentanyl and ventilated using the above algorithm. PACO2 was set at 35 torr. After 10 minutes of controlled ventilation NaHCO3 was infused at 4.4 mEq/min for 10 minutes. The procedure was then repeated a second time.

During the control period PaCO2 was 36.6 ± 2.8 as determined by arterial blood gas analysis. During NaHCO3 infusion PaCO2 increased to 38.8 ± 5.0 torr with a 59% increase in V̇CO2 and a 39% increase in V̇E. With the NaHCO3 infusion off, PaCO2 returned to 35.7 ± 3.9. During the procedure the mean end tidal CO2 concentration remained between 28.8 and 32.6 torr. The system maintained PaCO2 within 3.8 torr of the set point despite extreme changes in V̇CO2. Because a microprocessor was used for this application, the system is able to integrate multiple input signals for control of ventilation, set alarms and trend and print pertinent data.

II. Automated Fluid Infusion

The management of fluid resuscitation following burns and major trauma remains a difficult clinical area in which empirical judgement plays a major role. The measurement of urine output is often relied upon as a parameter of adequate perfusion and vascular volume repletion (3,4), however, from a practical standpoint, monitoring of urine flow and appropriately adjusting administered fluid is a problem. We have developed a system to automatically regulate the infusion of IV fluids to the burn/trauma victim to maintain a constant urine output rate. The patient's actual urine output rate is compared with the desired rate to generate an input signal to a digital process controller. IV fluids are administered to the patient at the calculated rate through a computer-controlled pump.

The purpose of the system is to facilitate routine fluid management; to provide accurate data on urine output and to improve fluid resuscitation during initial treatment period of shock and major burns.

Urine output is measured by a drop count technique. An infrared source and detector are positioned along a modified IV drop chamber to monitor urine as it flows from the urinary catheter. The infrared beam intensity is adjusted continually to compensate for chamber clouding or fluid accumulation on the wall. A similar technique is used for measuring the fluid infusion rate. The drop counter signals are processed by a microprocessor for calculating volumetric flow rate of fluid infused and urine output. An image CRT screen displays actual fluid infusion and urine flow in addition to cumulative fluid data for 1, 8, 24 hours post-burn.

The Intel 8085 microprocessor-based control system is shown in Figure 3. The microprocessor control over the IV infusion pump is

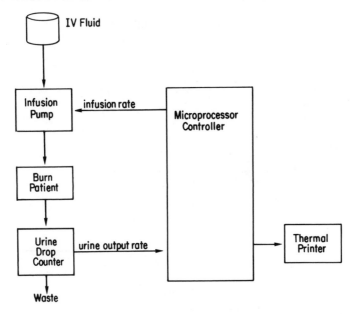

Figure 3.

via a digital interface. The fluid control algorithm, still under development, is designed to maintain urine output rate within the range .42-.85 cc/kg/hr. Fluid is infused initially according to the Parkland formula (5), the infusion rate is then modified depending on urine flow rates. Limits and safety features include maximum and minimum infusion rates.

The system described implements closed loop control of fluid infusion using only urine output. As maintenance of urine output is an important end point for manually administered fluid therapy, the computer controlled system is based around that parameter. Certainly cardiac parameters are important. However, in many patients, in whom invasive monitoring is not indicated, we believe that control of fluid infusion can be based on urine output alone.

III. Metabolism Monitoring

Using indirect calorimetry it is possible to determine the type of food-stuff metabolized by measuring the intake of oxygen, the output of carbon dioxide, and the excretion of nitrogen. (6) The total number of grams of protein oxidized is 6.25 times the nitrogen excreted. The oxygen and carbon dioxide involved in protein metabolism are subtracted from the totals, leaving the non-protein ($\dot{V}CO_2$) and ($\dot{V}O_2$). One can then apportion the relative contribution of the remaining oxygen and carbon dioxide using the respiratory quotient (RQ) of 0.71 and 1.0 for fat and carbohydrate, respectively. (7,8) Inherent limitations in the present methods of measuring

BLOCK DIAGRAM REPLENISHMENT TECHNIQUE

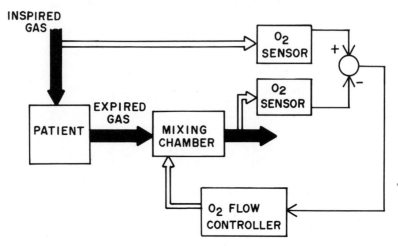

Figure 4.

$\dot{V}O2$, $\dot{V}CO2$ and RQ have made routine study of these patient metabolic parameters difficult. The instrument presented here enables the measurement of $\dot{V}O2$, $\dot{V}CO2$ and RQ simultaneously and continuously. The method incorporates a microprocessor controller and the gas replenishment feedback technique for both oxygen and CO2 measurements.

BLOCK DIAGRAM ABSORBTION – TITRATION TECHNIQUE

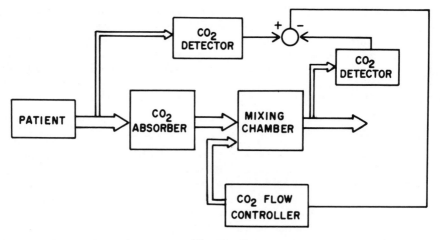

Figure 5.

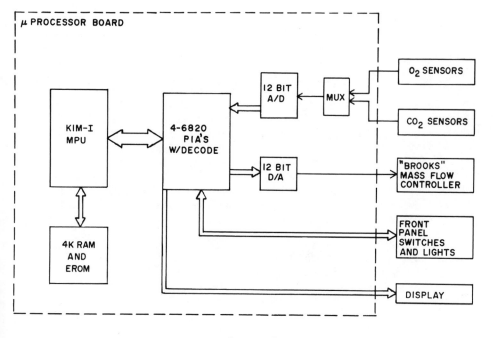

Figure 6.

$\dot{V}O2$ is measured using the technique shown in Figure 4. Oxygen is mixed with the patient's exhaled air until the mixture has the same oxygen concentratoin as the inspired air. The mixing proceeds continuously as the exhaled air passes through a mixing chamber. Feedback control is used to vary the inlet flow of oxygen until the mixing chamber effluent has the same oxygen fraction as the inspired gas. Once a steady state is reached, the oxygen replenishment rate is a continuous measure of the patient's mean $\dot{V}O2$. The $\pm$ 6% accuracy of the $\dot{V}O2$ measurement has been demonstrated in dog and human studies. (1,9)

Carbon dioxide production is measured (10) as shown in Figure 5. The CO2 fraction of the patient's exhaled air is measured by a CO2 detector. All of the carbon dioxide is then removed by passing the exhaled gas through a CO2 absorption chamber. In a subsequent chamber, pure carbon dioxide is metered into the chamber and mixed with the exhaled air until the CO2 fraction of the mixed gases is the same as that measured on the exhaled air. The flow of CO2 metered into the mixing chamber is a measure of the amount of CO2 produced by the patient. Sufficient accuracy of the CO2 monitor can be obtained without the need for absolute calibration of the CO2 detectors because they are nulled together to provide a relative difference rather than an absolute difference. Accuracy of the VCO2 measurement has been shown in an vitro experiment where the average percent difference between a simulated $\dot{V}CO2$ and the prototype's VCO2 measurement was 3.22%.

A MOSTECH 6502 microprocessor is used to control the two measurement systems and to display the data. Figure 6 is a block diagram of the system. Output displays show $\dot{V}O2$, $\dot{V}CO2$, metabolic rate (Cal/min) and the nutritive substrates in terms of gm/min of carbohydrate, fat and protein. This information will be used clinically to assess the relative efficacy of various modalities of nutritional support currently in widespread use for sustaining critically ill and injured patients.

REFERENCES

1. WESTENSKOW DR, RAEMER DB, GEHMLICH DK, JOHNSON CC: Instrumentation for monitoring oxygen consumption using a replenishment technique. J Bioengineering 2:219-227, 1978

2. MITAMURA Y, MIKAMI T, SUGAWARA H. YOSHIMOTO C: An optimally controlled respirator. IEEE Trans Biomed Engineering 18:330-337, 1971.

3. COPE O, MOOREFD: The redistribution of body water and the fluid therapy of the burned patient. Ann Surg 126:1010-1045, 1947

4. MOYLAN JA, et al: Postburn shock: A critical evaluation of resuscitation. J. Trauma 11:36-43, 1971

5. BOSWICHJA (ed): "Symposium on burns", The Surgical Clinics of North America, Volume 59, December 1978. Philadelphia, Pennsylvania: W.B. Saunders(s) Co

6. MAGNUS-LEVY A: The physiology of metabolism. In Von Noorden C (ed): Metabolism and Practical Medicine. London: William Heinemann, Volume 1, 1907, pp 185-204

7. HUNT LM: An analytic formula to instantaneously determine total metabolic rate for the human system. J Appl Physiol 27 (5): 731-733, 1969

8. BURSZTEIN, S, SAPHAR P, GLASER P,TAITELMAN U, et al: Determination of energy metabolism from respiratory functions alone. J Appl Physiol 42 (1): 117-119, 1977

9. CHEUNG P.W.,JOHNSON CC, MARTIN WE: Development of a continuous oxygen consumption monitor: The oxiconsumeter. Proc 26th Annual Conference of Engineers in Medicine and Biology, p 369, September, 1973

10. RAEMER DB, WESTENSKOW DR, GEHMLICH DK: Dynamic metabolic monitor: Continuous measurement of $\dot{V}O2$, $\dot{V}CO2$ and RQ. Proc 31st Annual Conference of Engineers in Medicine and Biology, p 235, October, 1978

RETROSPECTIVE AND PROSPECTIVE STUDIES OF A COMPUTERIZED ALGORITHM

FOR PREDICTING OUTCOME IN ACUTE POSTOPERATIVE CIRCULATORY FAILURE

William C. Shoemaker, M.D.

Departments of Surgery
Los Angeles County Harbor-UCLA Medical Center
Torrance, California

UCLA School of Medicine
Los Angeles, California

This project was supported, in part, by Grant Number HS 01833 from the National Center for Health Services Research, HRA.

ALGORITHM FOR OUTCOME PREDICTION

Therapy for postoperative shock is usually determined after identifying anatomical and physiological defects, which are then corrected. Cardiorespiratory monitoring usually is started after the patient is recognized to be in shock or in a critical condition with potentially lethal complications. The search for defects after they have developed is inefficient because it is a piecemeal approach that leads to fragmented episodic care; it is predicated on the concept of restoring normal values, and it is not based on the natural history of the shock states. More importantly, it does not provide a conceptual mechanistic framework for evaluating important physiologic interactions of volume, flow, oxygen transport, and tissue perfusion (1).

Previous studies showed that the survivors and nonsurvivors of life-threatening surgical conditions had markedly different cardiorespiratory patterns, despite a wide variety of surgical conditions and an even wider variety of surgical operations (2). In previous attempts to develop and outcome predictor, we used a method based on range criteria (3), which was a modification of the distribution free-method of Kendall and Stuart (4), a method using arbitrarily determined cut-points (5), and finally, the present algorithm (1).

The present article describes an algorithm to predict outcome based on retrospective analysis of 113 critically ill postoperative patients and this algorithm was tested prospectively on a new series of 60 patients. The algorithm quantifies the distance of observed values of each cardiorespiratory variable from empirically determined classification points that maximally separate the survivors' and nonsurvivors' values. An overall predictive index is determined as a weighted average of the distances for all variables. This algorithm was programmed for use on the hospital's computer and has been used over the past year as an objective heuristic method to evaluate complex physiologic interactions, to indicate the extent of each physiologic deficit in terms of its association with death or survival, and to evaluate the relative effectiveness of each therapeutic intervention in each patient (1,6).

The basic hypothesis is that physiologic changes, which are largely independent of the specific diagnosis or the surgical operation, are associated with survival or death and that the differences in the cardiorespiratory patterns of susrvivors and nonsurvivors can be used to predict outcome.

MATERIALS AND METHODS

Clinical Material

Patients were selected for monitoring on the basis of the severity of their illness and likelihood that this monitoring would be helpful in clinical decision-making. The clinical materials, techniques and methods for measurements of cardiorespiratory variables previously were described in detail (1,2,5).

Retrospective Analyses

Studies were conducted on 113 consecutively monitored patients who had undergone surgery for life-threatening illnesses. The patients were grouped as either survivors, rigorously defined as those who were eventually discharged from the hospital, and nonsurvivors, defined as those who died during their hospital stay (1,5).

Prospective Studies

After the predictive algorithm was formalized, it was tested prospectively in a new series of 60 consecutively monitored postoperative patients who sustained life-threatening surgical illnesses; 45 (75%) lived, 15 (25%) died. The last available set of measurements were compared with the actual final outcome to test the accuracy of the predictor.

Techniques for Measurement of Cardiorespiratory Variables

Analytical methods used for these studies were previously reported (1,2). Arterial pressures and cardiac output were measured using plastic catheters inserted into the femoral or axillary artery by percutaneous puncture or into the radial artery by cut-down. Central venous catheters and Swan-Ganz pulmonary arterial catheters were routinely placed by percutaneous technique. Intravascular pressures were measured continuously with transducers and recorded on an Offner recorder. The indicator dilution technique with indocyanine green was used for estimation of cardiac output; these were validated with direct Fick measurements of oxygen consumption and cardiac output.

Arterial and mixed venous (pulmonary arterial or right ventricular) blood samples were obtained anaerobically and analyzed for pH and blood gases at the time of cardiac output determination, i.e., within 30 seconds of the latter; rectal temperatures, hemoglobin and hematocrit values also were measured. Arterial and mixed venous oxygen saturation values were validated by direct measurements with a tonometer or cooximeter. Plasma volumes were measured with 125 I labeled albumin and blood volumes calculated from the latter and the simultaneously measured hematocrit.

The flow and volume variables were indexed according to the patient's body surface area. The data analyzed in this study were taken during periods remote from therapy, i.e., before therapy was given or after the immediate direct effects of the therapy has dissipated.

Criteria for Separation of Data into Stages

Some patients go through episodes of postoperative shock or circulatory instability more rapidly than others. For this reason, objective criteria for staging were devised so that each stage represents a comparable period of the shock syndrome (2). The preoperative period was defined as the control period (Stage A). Stage B was defined as the initial period of falling arterial pressure immediately after the onset of the etiologic event. The lowest recorded arterial pressure (Stage Low) separated Stage B from the middle period, Stage C. The latter was divided into Stages Cl and C2 by the point in time when arterial pressure returned halfway to the patient's control value. Stage D was defined as the late period for surviving patients when arterial pressures were normal or nearly normal. Stage E was defined as the preterminal period in patients who subsequently died; it covered the time from Stage C to about one or two hours prior to death. The final, agonal, terminal event was defined as Stage F(2).

Determination of the predictive (severity) index

From the data obtained in each successive stage of the tempo-
ral course of shock syndrome, a severity score was determined for
each cardiorespiratory variable (1). These scores then were combined into a single overall predictive index representing the entire set of cardiorespiratory variables. The algorithms for the
determination of severity score for each individual variable and
for the overall predictive (severity) index are described below.

Algorithm for Predictive (Severity) Score for Each Cardiorespiratory Variable

For each cardiorespiratory variable obtained within each shock
stage, two frequency distributions were determined, one for the
survivors, one for the nonsurvivors. Three regions were determined
from these two frequency distributions: (a) in the middle was the
overlapping region which contained data from both survivors and
nonsurvivors; (b) on one side of the overlap region was the survival region which included only 10% of the nonsurvivors' data and
the majority of the data obtained from the patients who lived; and
on the other side was the lethal region which contained only 10% of
the survivors' data and a majority of the data of the nonsurvivors
(Fig. 1). Within the overlap region a classification point, Q,
also was determined such that if this point were used to classify
the patients as survivors and nonsurvivors, it would yield the highest proportion of correct classification(7).

If NS and ND are numbers of survivors and nonsurvivors, RS and
RD are the numbers of survivors and nonsurvivors whose status were
correctly determined by a classification point, then the point Q
was determined so that RS/NS + RD/ND was a maximum (1,7).

For a given patient, the predictive score for each variable
was determined by comparing the observed value for each variable to
these three regions and the classification point, Q. It was determined such that the predictive score is negative if the measurements fall between Q and the fatal region and reaches −1.0 if the
measurement falls into the fatal region; similarly, it is positive
if the measurement is between the classification point, Q, and the
survival region and reaches 1.0 when the measurements fall in the
survival region. Specifically, the scores are determined by using
the following algorithm.

Let S10 and S90 be the 10th and the 90th percentiles of the data
obtained from the survivors and N10 and N90 be the 10th and the
90th percentiles of the data obtained from the nonsurvivors.

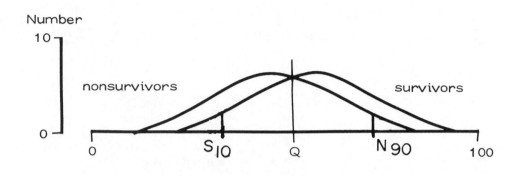

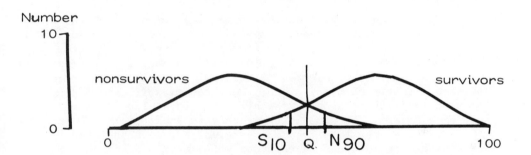

Figure 1. Schematic representation of the frequency distribution of survivors'and nonsurvivors' values of a cardiorespiratory variable, showing the Q
point which maximally separates the survivors from nonsurvivors, the N90 the 90th percentile of nonsurvivors,
and the S10, the 10th percentile of survivors. Upper
diagram shows minimal separation of the survivor and
nonsurvivor distributions giving poor predictive capacity. Lower diagram shows wide separation of the distributions of survivors and nonsurvivors, indicating
good predictive capacity of this variable.

Denote the observed value of a variable by x and the value of the severity score for this variable by P. When the survival region is to the right of the overlap region:

$$P = -1, \text{ if } x \text{ is less than } S_{10}$$

$$P = \frac{x - Q}{Q - S_{10}}, \text{ if } x \text{ is between } S_{10} \text{ and } Q$$

$$P = \frac{x - Q}{N_{90} - Q}, \text{ if } x \text{ is between } Q \text{ and } N_{90}$$

$$P = +1 \text{ if } x \text{ is greater than } N_{90}$$

When survival region is to the left of the overlap region, the formulas were similar except that S10 is replaced by N10 and N90 is replaced by S90, and all equations are multiplied by -1.

Determination of the Overall Predictive (Severity) Index

Within each stage of shock, an overall predictive index was calculated as the weighted average of the predictive scores for all individual cardiorespiratory variables. The weight assigned to a particular cardiorespiratory variable is its accuracy in predicting the actual outcome (Table 1). Specifically, it is the average proportion of correct predictions; when the classification point, Q, is used for each available variable, it is calculated as: 1/2 (RS/NS + RD/ND). The predictive index, in an approximate sense, is interpreted as the likelihood of survival based on the overall cardiorespiratory status in patients with shock of the same etiology and stage (1,7). A value of +1 indicates maximal likelihood of survival, a value of -1 indicates maximal likelihood of nonsurvival, and values between these indicate lesser degrees of certainty in predicting outcome, with a value of zero indicating equal likelihood of survival and nonsurvival.

RESULTS

Evaluation of the Predictive (Severity) Index

The predictive index was evaluated by examining each of the cardiorespiratory values of the 113 postoperative patients within each stage of the shock syndrome. The predictive indices for each stage of the survivors and nonsurvivors are given in Table 2. The mean predictive index of the survivors significantly differed from that of the nonsurvivors for all stages.

TABLE 1

CLASSIFICATION POINT (Q) FOR EACH VARIABLE AT EACH STAGE

STAGES

Variables	B Q	B WT	LOW Q	LOW WT	C_1 Q	C_1 WT	C_2 Q	C_2 WT	LATE Q	LATE WT
Mean arterial pressure	>72.9	.51	>61	.49	>71.2	.38	>85.5	.40	>71.9	.82
Central venous pressure	∧∨7.2	.29	>2.75	.23	∨0.527	.16	∨5.13	.22	∧∨4.98	.31
Central blood volume	∧∨707	.23	∨616	.25	>915	.10	>1272	.33	∨∨1126	.19
Stroke index	∧∨43.2	.20	>27.1	.33	>26.4	.35	>37.1	.22	>36.7	.59
Hemoglobin concentration	∧∨8.7	.46	>8.7	.38	>8.68	.20	>8.77	.27	>10.1	.25
Mean pulmonary arterial pressure	∨16.7	.35	∧14.5	.20	∨15.8	.23	>18.3	.47	>13.2	.50
Pulmonary wedge pressure	∧∧5.13	.52	>8.8	.38	>7.78	.28	>4.87	.29	>1.96	.50
Red cell mass deficit or excess	∧∧395	.66	∧∧-396	1.0	>-422	.47	>-312	.48	∧-310	.86
Blood volume deficit or excess	∧-660	.37	∧-557		>-684	.42	>-867		116	1.0
Cardiac index	∧3.18	.34	∧2.65	.42	>3.34	.42	>4.35	.27	>3.55	.63
Mean transit time	∧∨19.7	.42	>21.3	.47	∨13.4	.36	∨∧3.4	.36	∨9.4	.54
Left ventricular stroke work	∧∨35.3	.36	>27.7	.52	>39.8	.31	∧47.8	.31	∨38.5	.74
Left cardiac work	∧3.28	.46	∧2.32	.53	>4.07	.43	∧4.63	.35	>3.97	.79
Mean systolic ejection rate	∧203	.40	∧129	.30	>226	.48	>309	.28	>280	.37
Tension time index	∧133	.30	>697	.60	>893	.23	∨1214	.22	>792	.78
Right ventricular stroke work	∧5.3	.24	>6.0	.25	∨4.96	.43	∨1.3	.52	∨12.6	.63
Right cardiac work	∧0.457	.24	∧∧.496	.35	∨0.604	.35	∨0.987	.47	∧0.848	.38
Systemic vascular resistance	>2087	.26	∧1460	.21	∨2193	.40	∧2551	.15	∧1179	.36
Pulmonary vascular resistance	∨347	.75	>417	.53	∨271	.21	>406	.43	>603	.50
Heart rate	>78.5	.14	∨93	.21	∨99.5	.43	∧109	.16	∧101	.27
Temperature, rectal	∨98.9	.33	∨99.6	.33	>99	.18	>100.1	.16	>98.2	.33
Arterial pH	>7.39	.50	>7.39	.60	>7.33	.16	>7.45	.49	>7.40	.63
Arterial CO_2 tension	∨35.3	.47	∨39.3	.50	>28.4	.24	>30.9	.16	∨42	.53
Mixed venous O_2 tension	>25.6	.22	>31.6	.52	>33.4	.27	>29.1	.21	>30	.48
Arterial hemoglobin saturation	∨97	.46	∨98.5	.26	∨97.6	.21	>97	.30	>92.7	.48
Arteriovenous oxygen content diff.	∨5.39	.28	∨4.68	.43	∨3.88	.36	>3.32	.25	∧3.32	.67
Oxygen availability	>472	.67	>461	.55	>470	.47	>416	.20	>338	.62
Oxygen consumption	>126	.47	>100	.24	>151	.29	>161	.31	>118	.43
Oxygen extraction ratio	∨37.3	.28	∨27.4	.55	∨28.6	.33	∨23.2	.25	>33.9	.63
Red cell flow rate	∧1.02	.50	>1.15	.44	>0.994	.40	>1.27	.17	∧0.689	.86
Blood flow/volume ratio	∧1.12	.26	>1.41	.75	>1.30	.37	>1.49	.28	∧1.42	.86
Oxygen transport/red cell mass	∨0.171	.33	∨0.256	.67	∨0.170	.35	>0.222	.67	>0.192	.52
Tissue oxygen extraction index	∨7.2	.63	∨5.30	.60	>3.69	.53	>6.08	.21	∨4.80	.67
Efficiency of tissue oxygen extrac.	∨8.09	.83	∨9.86	1.0	>8.46	.60	>6.04	.67	>3.59	1.0
Oxygen transport/red cell flow	∨192	.50	∨180	.64	∨129	.38	∨183	.26	>116	.30

TABLE 2

MEAN PREDICTIVE INDEX VALUES OF SURVIVORS AND NONSURVIVORS

Stage of Shock	Survivors	Nonsurvivors	P Value*
B	+0.22±.33** (20)***	−0.23±.32 (14)	< .001
Low	+0.32±.42 (21)	−0.34±.28 (15)	< .001
C_1	+0.26±.39 (58)	−0.19±.32 (29)	< .001
C_2	+0.18±.25 (49)	−0.22±.27 (22)	< .001
Late	+0.42±.23 (29)	−0.55±.28 (13)	< .001

*P value obtained from STudent's t-test

**Mean⁺SD

***Number of patients

In order to examine the reliablility of the predictive index as an outcome predictor, the calculated predictions were compared with the actual outcomes of the first series of study patients. Table 3 shows the proportion of correct predictions when the patients were catergorized into four groups: those who survived, those who died directly as the result of shock, those who died of complications, and those who died from a subsequent unrelated etiologic event, such as a second operation. The proportion of correct predictions was highest during Stage Low and the Late Stages, but was adequate during other stages.

Prospective Evaluation of the Predictive Index

The predictive index was tested prospectively in 60 consecutive general surgical patients who had undergone operations for life-threatening conditions. When the predictions of the last available data were compared with the patient's actual outcome, the outcome of the 60 patients was correctly predicted by the algorithm in 54 (90%) patients and six (10%) of the cases were incorrectly predicted. Three (5%) of these were predicted to live; that is, their last predictor was in the positive range, but these patients went on to die. Two of these patients were operated upon with the intent of performing an extensive ablative carcinoma procedure, but at operation, advanced carcinomatosis was found and the patients were closed with total operative times of 60 and 90 minutes. One of these patients died on the 22nd day and the second patient died

TABLE 3

PERCENTAGE OF PATIENTS CORRECTLY PREDICTED BY STAGE

Group	Stage B % (No.Correct/ Total)	Stage LOW % (No. Correct/ Total)	Stage C$_1$ % (No. Correct/ Total)	Stage C$_2$ % (No. Correct/ Total)	Late Stages % (No. Correct/ Total)
Survivors (Sensitivity)*	70 (14/20)	81 (17/21)	78 (45/58)	78 (38/49)	93 (27/29)
Nonsurvivors (Specificity)**	86 (12/14)	87 (13/15)	76 (22/29)	86 (19/22)	92 (12/13)
Died of shock	(3/3)	(3/4)	(11/13)	(7/8)	(10/11)
Died of complications	(7/9)	(7/8)	(9/13)	(10/11)	(2/2)
Died of unrelated causes	(2/2)	(3/3)	(2/3)	(2/3)	(0/0)
Total correctly predicted	76 (26/34)	83 (30/36)	77 (67/87)	80 (57/71)	93 (39/42)

*   Sensitivity is defined as the proportion of survivors whose outcomes are correctly predicted.

** Specificity is defined as the proportion of nonsurvivors whose outcomes are correctly predicted.

11 days postoperatively, of advanced carcinomatosis. A third patient was monitored only during the first day postoperatively, and
after a stormy course died of an intra-abdominal abscess on the
22nd postoperative day. There were also three (5%) patients who
were predicted to die, but lived. All of these three patients were
monitored only during the day of the operation and the first postoperative day.

Several interpretations of the errors are obvious. First, the
predictor, which was based on cases observed during the previous
decade, may be inadequate for present day technology; i.e., at the
present time it may be possible to salvage some previously unsal-
vageable patients with newer and better techniques. Second, some
of the prospectively monitored patients may not have been observed
long enough for satisfactory predictions. For example, if the monitoring is stopped the day after surgery, the data should not be
expected to predict outcome in patients with late complications.
Moreover, the predictor measured before active aggressive therapy
inadequately assesses the circulatory status after that therapy.
Finally, a predictive index based on acute circulatory changes
should not be expected to predict outcome from noncirculatory problems, such as the progression of carcinomatosis.
Relative Usefulness of the Predictive Index

Of the 60 patients who were monitored, 13 (22%) had positive
predictors throughout, indicating survival. These patients all did
well, and probably would also have done well without monitoring.
That is, they may not have needed the monitoring except to document
that the patients were on a "survival" course, and as a prophylac-
tic measure to anticipate early or incipient problems.

Four patients (7%) had a progressive, inexorable downhill
course. This was indicated by negative predictive indices, which
became progessively more negative; therapy failed to reverse either the clinical course or the predictors; all of these patients
died despite aggressive and adequate therapy. One (2%) patient was
monitored in the terminal state; i.e., very late in the course of
postoperative shock; she also went onto a lethal outcome.

Twenty-six (43%) of the patients were strongly negative sug-
gesting a lethal outcome, but were brought into the positive range;
they survived as predicted by their latest available values. In
many of these instances the therapy was specifically based on the
observed values and their predictors; that is, the predictor was
used to titrate the patient to "optimal" physiologic values.

Three (5%) of the patients had minor improvement of their
predictors gradually over a period of days. These patients probably would have improved with time without the monitoring;
essentially, the monitoring was not used (or was not useful) and
did not contribute appreciably to the decision-making involved.

Seven (12%) of the patients transiently improved from strongly negative range. However, they subsequently returned to the negative range; thus, they were predicted to die, and indeed, did die despite aggressive and enthusiastic therapy.

DISCUSSION

When cardiorespiratory monitoring of critically ill patients is undertaken, the clinician is often confronted with an enormous mass of physiologic data. The overall status of the patient as indicated by these data may be difficult to discern, and trends of improvement or lack of it over time may be difficult to establish, particularly if apparently related variables change in different directions.

The algorithm developed from a previous group of critically ill patients when applied to a new group of patients was reasonably satisfactory in predicting outcome and established this procedure as a potentially useful approach. When refined and further developed with appropriate testing, it is anticipated that this approach will have greater usefulness as a measure of the process of care in patients with acute circulatory failure.

The present algorithm provides an index which predicts outcome by its sign and by its magnitude is a measure of the severity of illness. The reliability of this method was evaluated in the following manner: (a) as a predictor of outcome, the index provided a rather high sensitivity, specificity, and predictive accuracy; (b) used as a measure of severity, the mean index value for survivors was statistically significantly higher than that for nonsurvivors at each stage of shock; and (c) the algorithm was tested prospectively in a new series of patients with satisfactory results. This tends to support the basic hypothesis that the pattern of cardiorespiratory values is related to outcome. The proposed algorithm quantifies the probability of survival (or death) and, thus, provides a useful index to evaluate the circulatory status. In essence, the pattern of changes in the predictive index repeatedly measured at intervals provides a useful means of quantitatively assessing trends in overall cardiorespiratory status.

The present algorithm serves as a yardstick by which specific therapeutic interventions may be assessed in terms of improvement or lack of improvement in overall cardiorespiratory status. The effect of therapy is difficult to evaluate when the values of all variables are considered individually or when observed values are compared with normal values. Furthermore, when some variables change in one direction, while others change in the opposite direction, the overall effect is difficult to determine. However, the present algorithm which is based on a multivariate analysis that weighs each variable according to its predictive capacity, may be

used to quantify the cardiorespiratory status before, during, and
after each specific therapy. Improvement in the overall status of
the patient also will be reflected by a change of the predictive
index in the positive direction. Thus, it is possible to quantitate the effect of each therapeutic intervention on each cardiorespiratory variable as well as on the composite and on the overall
predictive index.

SUMMARY

    An index for prediction of outcome was developed by a nonpar-
ametric multivariate analysis of cardiorespiratory data from 113
critically ill postoperative general surgical patients and was
tested prospectively in a new series of 60 patients. This predictive index was based on a computerized algorithm that compares a
given observed value for each cardiorespiratory variable with the
frequency distributions of survivors' and nonsurvivors' values of
that variable; then an overall predictive index was developed from
the sum of the predictive score of each variable weighted according
to its predictive capacity. The difference in the mean values of
the predictive index for survivors and nonsurvivors was statistically significant (P .001) during each stage of shock. Sensitivity
of the index in prediction of survival ranged from 70% to 93% depending upon stage, the specificity of the index ranged from 76% to
92%, and the predictive accuracy ranged from 87% to 96%. There was
90% accuracy when the algorithm was applied prospectively to a new
group of patients.

    The usefulness of this index in tracking the course of criti-
cal illness was demonstrated in the prospective series. We conclude that the predictive index greatly aids in the interpretation
of monitored cardiorespiratory variables, facilitates the diagnosis
of physiologic impairments, and clinical decision-making at the
bedside.

REFERENCES

1. SHOEMAKER, W.C., PIERCHALA, C., CHANG, P., CZER, L., BLAND, R. SHABOT, M.M., AND STATE, D.: Cardiorespiratory monitoring in postoperative patients: I. Prediction of outcome and severity of illness. Crit Care Med 7:236-241, 1979.

2. SHOEMAKER, W.C., MONTGOMERY, E.S., KAPLAN, E. AND ELWYN, D.H.: Physiologic patterns in surviving and nonsurviving shock patients. Arch Surg 106:630-636, 1973.

3. SHOEMAKER, W.C., ELWYN, D.H., LEVIN, H., AND ROSEN, A.L.: Use of nonparametric analysis of cardiorespiratory variables as early predictors of death and survival in postoperative patients. J Surg Res 17:301-314, 1974.

4. KENDALL, M.G. AND STUART, A.: Advanced Theory of Statistics. Vol. 3. New York, Hofner Publishing Co., Second Ed, 1968.

5. SHOEMAKER, W.C., PIERCHALA, C., CHANG, P., AND STATE, D.: Prediction of outcome and severity of illness by analysis of the frequency distributions of cardiorespiratory variables. Crit Care Med 5:82-88, 1977.

6. SHABOT, M.M., BLAND, R., SHOEMAKER, W.C., AND STATE, D.: On-line computerized cardiorespiratory analysis and predictive indices for critically ill surgical patients. Arch Surg. Submitted.

7. PIERCHALA, C., SHOEMAKER, W.C., AND CHANG, P.: Comparison of some methods for early classification of survivors and nonsurvivors of postoperative shock. Computers Biol Med 8:279-292, 1978.

8. BLAND, R., SHOEMAKER, W.C., AND SHABOT, M.M.: Physiologic monitoring goals for the critically ill patient. Surg Gynecol Obstet 147:833-841, 1978.

CLINICAL EVALUATION OF THREE TYPES OF CONTINUOUSLY MEASURING OXYGEN SENSORS - THEIR POTENTIAL USE FOR COMPUTER CONTROL OF BLOOD OXYGENATION

Edwin G. Brown, Francis E. McDonnell,
Eresvita E. Cabatu, Roy A. Jurado,*
Susan L. Maier, and Robert J. Wynn

Department of Pediatrics
The Mount Sinai School of Medicine of
the City University of New York
New York, New York 10029

*Division of Cardio-Thoracic Surgery
Department of Surgery
The Mount Sinai School of Medicine of
the City Univeristy of New York
New York, New York 10029

ACKNOWLEDGEMENTS

The authors wish to acknowledge the help of the medical and nursing staff in the Neonatal and Cardiothoracic Intensive Care Units at The Mount Sinai Medical Center, New York for their patience and help in obtaining the data used to evaluate the sensors. We are indebted to Mrs. Georgia Coar for typing the manuscript and Avron Y. Sweet and Richard W. Krouskop for their helpful suggestions in the preparation of the manuscript.

This research was supported in part by the National Heart, Lung and Blood Institute, National Institutes of Health, Bethesda, Maryland (Contract No. 1 HR 5-2949) The National Foundation - March of Dimes and H.E.W. Maternal and Child Health Training Grant No. MCT-002018

In critical care patients, many vital parameters are continuously measured, recorded and graphically displayed with the aid of a computer. However, important measurements such as oxygen partial pressure (PO2) and hemoglobin saturation (SO2) usually are measured in laboratory analyzers from blood samples withdrawn intermittently through a cathether. These intermittent measurements provide values only for when the blood is sampled and dynamic changes cannot be shown. Continuous monitoring of PO2 and/or SO2 is desirable to provide for better control of oxygenation. Decreased frequency of blood sampling will reduce the opportunity for contamination of the catheter and eliminate the danger of diminished blood volume especially in the neonate or very small child.

If there are sensors that can continuously measure PO2 and SO2 accurately and reliably, the data can be computer processed and used to monitor patient care and possibly servo-regulate the fractional inspired oxygen (FiO2). Three systems which show promise are undergoing laboratory and clinical testing in the neonatal intensive care and cardiothoracic units at The Mount Sinai Medical Center. Enough data have been obtained to determine their accuracy and reliability. The systems include: 1. a polarographic PO2 catheter sensor - Brown and McDonnell, 2. an oximeter SO2 catheter sensor - Oximetrix Inc., Mountain View, California and 3. cutaneous PO2 sensors - Roche Medical Electronics Division, Cranbury, New Jersey and Radiometer, London Company, Cleveland, Ohio. A description of each system, its functional characteristics and the possible applications for patient monitoring are the subject of this report.

METHODS, PROCEDURES AND RESULTS

1. CATHETER PO2 SENSOR

This sensor measures PO2 based on the polarographic determination of oxygen. (1) The operating principles of polarographic sensors are well documented. (2) The system consists of a catheter which contains a gold wire cathode and a sintered silver-silver chloride anode. The cathode is incorporated into the wall during the extrusion of a 3 or 5 French polyurethane catheter; the anode is located in the eletrolyte chamber in the catheter Luer connector (Fig. 1). Because the sensor is in the catheter wall, the lumen is unobstructed and makes possible blood sampling, pressure measurements and fluid infusion. The density of the gold wire provides a radio-opaque line and permits the position of the catheter to be localized on the roentgenogram. It is designed to be disposed of after single use.

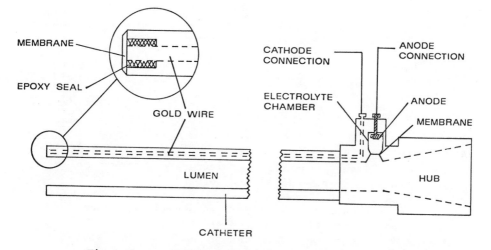

Fig. 1   The oxygen sensing catheter electrode is shown in diagramatic form. The gold cathode is sealed and bonded to the catheter by an epoxy resin. A cellulose acetate membrane is formed over the cathode by solvent evaporation. The reference electrode is a sintered silver-silver chloride pellet located in the electrolyte chamber of the Luer hub. A micro-porous membrane separates the anode from the hub lumen. The electrolyte chamber is filled with electrolyte through an injection port not shown in the diagram. The anode and cathode are coupled to the body electrolyte when the catheter and hub lumen are filled with an electrolyte solution.

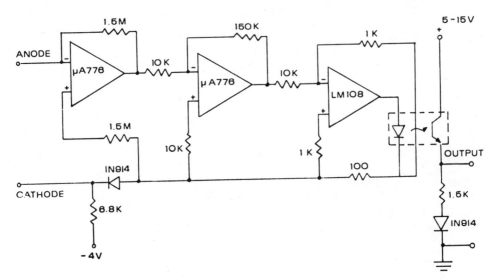

Fig. 2 Circuit diagram of the electronics used to operate the catheter sensor. An integrated circuit operational amplifier uA776 with a forward biased silicone diode 1N914 provides a constant polarizing voltage and acts as a current amplifier. An additional operational amplifier LM108 is used as the second stage to drive the optical isolater. The output from the optical isolator can be used to drive a metered or digital display device and a recorder.

The anode and cathode are connected to a pre-amplifier (Fig. 2) which can provide a constant negative voltage between 0.6 and 0.9 volts and detect currents on the order of 10 -9 A. The electrodes are isolated from other electrical apparatus by means of an optical coupling device. The pre-amplifier drives a data processor which has digital and analog displays. The processor has controls for calibrating the sensor manually, but it also can be programmed to automatically calibrate the sensor during use. (3)

Sensor Performance In Vitro

These tests were performed in a specially constructed chamber in which the temperature and the rate of flow of the test solution could be accurately controlled. The solutions used were plasma or physiologic saline. Catheter electrodes were connected to the polarizing voltage and allowed to stabilize for 30-60 minutes. During stabilization there was a rapid non-linear decline in the sensor current. The average drift following stabilization was less than 1 percent per hour. Many of the sensors required no calibration during a 72 hour test period. The range of current output of the electrodes was between 30 and 60 x 10 -9 A. In an air equilibrated 37°C saline solution (147 mmHg PO2) at a flow velocity of 10 cm per second. The sensors were linear over a PO2 range of 0 to 500 mmHg and for practical purposes, free of hysteresis. They were sensitive to flow when the flow velocity was varied from 0 to 4 cm per second and the current varied as the log of the flow velocity. The output current in saline was constant while the flow velocity was varied over the range of 4 to 14 cm per second. The flow velocity of blood in the major arteries of adults and infants is fast enough to prevent any influence from the boundary layer that would form in the blood adjacent to the sensor. The temperature coefficient was 2 percent per degree centigrade at 37°C. The 95 percent response time to a step change of PO2 from 84 to 147 mmHg was less than one second and there was no evidence of hysteresis when the PO2 was returned to 84 mmHg. The sensor current increased when PCO2 was increased in an unbuffered saline test solution. The change appeared to be linear and was 0.2 percent per mmHg of PCO2 at a PO2 of 147 mmHg. However, when the PCO2 was varied in the same test solution buffered to maintain the pH between 6 and 8, there was no significant change in the sensor current.

Sensor Performance in Baboons

Sensors inserted into the iliac artery of baboons for three to seven days were evaluated for function and biocompatibility. (4)

Before and during implantation, and after the sensor was removed, extensive analysis of blood samples for biochemical, hematological and coagulation abnormalities was performed. There were no abnormalities found during the study or several months later. The electrode structure was not damaged and there were no clots on the sensor as observed by light and scanning electron microscopy; platelets adhered to the cellulose acetate membrane on some of the sensors. Studies using the animal's own kidney for a micro-embolus trap done according to the method of Kusserow, were negative. (5) There were no clots in the large vessels near the location of the catheter. Arterial blood pressure was measured continuously through the catheter lumen by a Statham P 23 series pressure transducer (Statham Medical Instruments, Inc., Hato Rey, Puerto Rico). The transducer did not cause electrical interference with the sensor current and the pressure wave form recordings were of excellent quality throughout each study. A 60 mmHg change in systolic blood pressure did not affect the sensor current values. Changes of plasma osmolality over a range of 260 to 350 mOs per kg of water also did not significantly change the sensor current. Sixty to ninety minutes were allowed for the sensor to stabilize; during the ensuing six hours the output current varied no more than 10 percent from the arterial PO2 (PaO2). One point calibration was possible over a six hour period and the sensor provided useful PO2 measurements for at least seven days. The sensor appeared to be non-toxic to the metabolic, hematologic and reticuloendothelial systems of the baboon.

Sensor Performance in Humans

A catheter sensor coupled to a silver-silver chloride anode as shown in Figure 1 was inserted into the umbilical artery of five infants and advanced into the descending aorta, just above the diaphragm. The position of the catheter was verified by x-ray. One sensor was inserted into the umbilical vein and advanced into the inferior vena cava. Under direct visualization, catheter sensors were inserted through an introducer into a femoral artery and advanced into the aorta above the bifurcation in three adult cardiothoracic patients. Electrolyte was injected into the anode chamber through a side port (not shown in Fig. 1). A solution of electrolyte in 5 or 10 percent dextrose was infused at a rate of 3 ml per hour through the catheter lumen by interposing a Sorensen flush valve (Sorensen Research Company, Salt Lake City, Utah) between the catheter and an infusion pump, except in two infants in whom a heparin lock was used. The sensor was connected to the electronics by a coupler attached to the Luer hub connection.

The infants at birth were 32 to 40 weeks gestation (mean 37 weeks) and they weighed from 1800 to 4030 gm (mean 2882 gm.) They had a variety of illnesses including asphyxia neonatorum, transient respiratory distress and hyaline membrane disease. Two of the adult cardiothoracic patients had coronary by-pass operations and the third received a mitral valve prosthesis. The clinical evaluation of the catheter sensors was approved by the Human Investigation Committee of The Mount Sinai Medical Center. Informed consent was obtained from the adult patients and from the infants' parents.

Of those sensors inserted into the vessels of infants and adults, two did not function properly; the current output did not correlate with the PO2 values from the laboratory analyzer. An erratic sensor current was noted while the laboratory PO2 values were stable. No clear cause was found for the failure of these sensors, but a heparin lock was used in them. The remainder of the sensors functioned well and accurately represented the PO2 after a 30 to 60 minute stabilization period. In order to determine the drift characteristics, the sensors were not calibrated after the initial PO2 value was obtained. Blood samples were taken when necessary as judged by the medical staff without regard to the output current of the sensor. By the least squares method, a linear regression of simultaneous sensor output current compared to a corresponding laboratory PO2 value was established for each sensor except for the two that failed to function from the beginning. For each of the seven patients whose data could be analyzed, a correlation coeffi-

cient was determined. This analysis was done on all of the data
pairs obtained during the time the sensor was used (Table 1).
Spurious readings from the processor display due to electrical interference and occasional inaccurate laboratory blood gas determinations adversely affected the correlation coefficient of sensors
Kr and Bl. Despite these problems and the lack of regular calibration, the output current correlated satisfactorily with the laboratory analysis. Accuracy within 10 percent of the laboratory value
would be possible if the sensors were calibrated every six hours
during the first 24 hours. This was determined indirectly from the
in vitro and baboon studies, and from four patients who had a sufficient number of PO2 determinations to permit evaluation of the
hourly output current drift. After the first 24 hours, the sensors
remained responsive to changes in PO2 and were remarkably stable
with a drift of less than 0.5 percent per hour.

The offset or ohmic current normally present in polargraphic
sensors did not change appreciably. This, together with the drift
characteristics, indicate that the sensor is definitely capable of
one point calibration. For the best possible results, calibration
should be performed 60 minutes after polarization and every six
hours during the first 24 hours and one or two times during each 24
hour period thereafter.

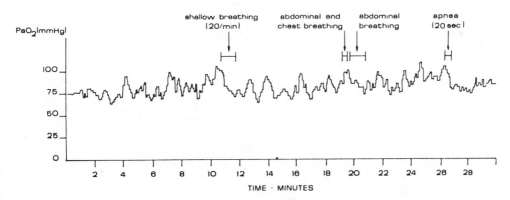

Fig. 3 PaO2 values of an infant with variable and
rapidly changing breathing patterns who was in an FiO2
of 0.30 throughout. Some of the respiratory patterns
are indicated. The 30 minute segment shown is illus-
trative of the widely fluctuating PaO2 evident during
the many hours of monitoring.

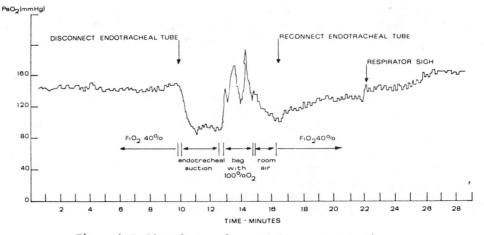

Fig. 4 PaO2 values of an adult postoperative cardiothoracic patient who required assisted ventilation.
The PaO2 is stable except during the indicated therapeutic manipulations. It should be noted that after
these procedures the PaO2 recovered slowly and did not
attain the original concentration until 14 minutes
later.

The device can accurately measure venous PO2 as shown by the
correlation coefficient of Ma in Table 1. It is sufficiently sensitive to detect small and abrupt changes in PO2. As shown in Figure 3, PaO2 changes in a newborn infant were readily and immediately detected following the onset of alteration in the breathing pattern. In infants, PaO2 is very sensitive to small changes in ventilation; adult patients show a greater stability of PaO2 (Fig.
4). Further, in Figure 4 are shown abrupt and sometimes marked
changes in PaO2 that occurred with therapeutic intervention.

Each sensor was carefully examined for clots and damage to the
component parts after it was removed form the patient. None were
found by light or scanning electron microscopy. No clotting defects or platelet destruction was observed while the sensors were
in use.

2. OXIMETER SO2 CATHETER SENSOR

This sensor measures hemoglobin saturation by the method of
reflectance oximetry. (6,7) A 4 French polyurethane catheter containing two plastic fiber optics and having an unobstructed central
lumen is used to transmit a light source to and from the blood

stream. The light source consists of three diodes that emit three different wave-lengths of red light. Each is pulsed in sequence at 1 msec intervals through one of the fiber optics. Light is reflected by the blood and transmitted to a light detector by the other fiber optic. The intensity of each wave-length of light is converted to an electrical signal by the detector and transmitted to a data processor. Using experimentally derived equations, the processor computes the average SO2 every second, stores the signals and displays the result in digital form for the preceding five seconds. The data can also be continuously displayed on a two speed recorder supplied with the processor. The processor can also determine if the intensity of the reflected light is different from that ordinarily measured in flowing blood. If it is different, an artifact condition alarm can alert the operator to possible problems which include: the catheter tip against a vessel wall, a disconnected or non-working light source, damaged fiber optics and a clot which covers the fiber optics on the catheter tip.

A unique part of this system is the method used to calibrate the sensor. The catheter is packaged by the manufacturer with its tip inserted into a disposable optical reference assembly. The fiber optics are connected to the light source and the catheter tip is moved against the reflector surface of the reference assembly. The oximeter measures the light intensity and within 30 seconds, the system is calibrated and ready for use. The sensor can also be calibrated while in the patient. This requires that a blood sample is withdrawn through the catheter lumen and an SO2 measured by a laboratory oximeter. The SO2 value on the processor display is then corrected for the difference between the laboratory value and the processor value at the time of sampling.

Sensor Performance in Humans

Seven catheters were inserted into the umbilical artery of six infants and advanced into the descending aorta just above the diaphragm. The position of the catheter was verified by x-ray. The system was calibrated before the catheter was inserted by the method described above. A solution of electrolyte in 5 or 10 percent dextrose was infused at a rate of 3 ml per hour through the catheter lumen. A total of 238 blood samples (0.4 ml) were obtained to compare the in vivo arterial SO2 (SaO2) with a laboratory SaO2 analysis. The laboratory analysis was done with a Radiometer OSM2 oximeter which requires as little as 0.02 ml of blood for each measurement. The OSM2 oximeter was calibrated every six to twelve hours using the patients own blood saturated to 100 percent in a tonometer. This method of calibration reduces to a minimum, errors that might be introduced by changes in hematocrit and bilirubin concentrations. The PaO2, PaCO2 and pH were also measured with a Corning 175 blood gas analyzer (Corning Medical, Medfield, Massachusetts). The infants at birth were 27 to 36 weeks gestation (mean 32 weeks) and weighed from 760 to 2420 gm (mean 1604 gm).

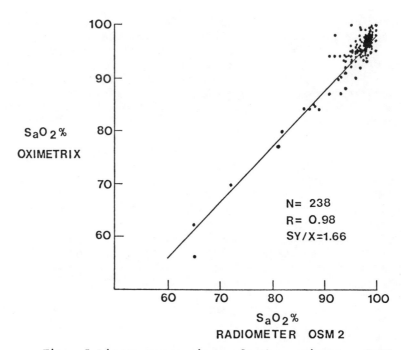

Fig. 5 Linear regression of the oximeter SaO2 catheter sensor and the SaO2 as measured by the Radiometer OSM2 laboratory oximeter. The correlation coefficient is 0.98 from 238 comparisons.

In a sensor whose performance was adequate, a clot formed on the tip twenty-four hours after it was inserted; this was indicated by the reflectance intensity alarm. Since blood could not be withdrawn through the catheter lumen, it was removed and another one inserted. The other sensors functioned well and accurately represented the SaO2 after they were inserted into the patient. The catheters remained in the patients from 13 to 155 hours (mean 71 hours). One attempt to insert an oximeter catheter was unsuccessful and efforts to pass a conventional catheter also failed.

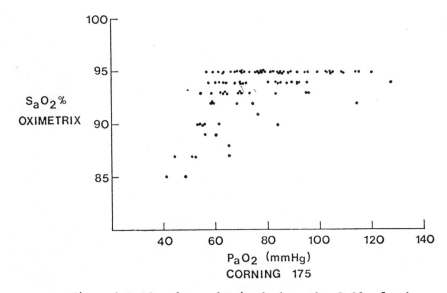

Fig. 6 PaO2 values obtained when the SaO2 of the oximeter catheter sensors varied between 85 and 95 percent. The minimum and maximum PaO2 values were 41 and 127 mmHg respectively.

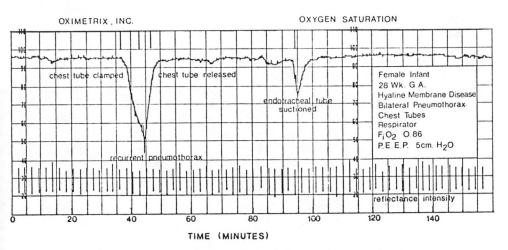

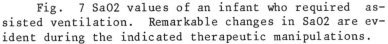

Fig. 7 SaO2 values of an infant who required assisted ventilation. Remarkable changes in SaO2 are evident during the indicated therapeutic manipulations.

A linear regression analysis of simultaneous catheter oximeter SaO2 compared to corresponding laboratory SaO2 values was done using all values which included 238 data pairs. The correlation coefficient of these data was 0.98 (Fig. 5). The accuracy of the instrument was not affected by changes in bilirubin or hematocrit concentration, type of hemoglobin (Hgb A, Hgh F), pH or PaCO2. As shown in Figure 6, the PaO2 varied from 41 to 127 mmHg when the SaO2 was between 85 and 95 percent. The catheter oximeter immediately detected changes in SaO2 during therapeutic interventions (Fig. 7).

3. CUTANEOUS PO2 SENSORS

The Roche sensor is a polarographic Clark-type electrode with a large size cathode (diameter 4 mm). (8) Since the surface area of the sensing electrode is larger than that of most other similar type of sensors,(9) the PO2 value is representive of a larger area of skin. This should improve the average value of cutaneous PO2 measurements. In order not to disturb the tissue oxygen profile when an electrode with a large sensing surface is employed, a membrane with low oxygen permeability must be used. The sensor current output becomes limited by oxygen diffusion through the membrane, thus it measures PO2 at the skin surface. The electrode is heated to 44°C to cause hyperemia. This increases the amount of oxygen that diffuses from the subcutaneous vascular bed to the skin surface. The sensor is attached to the skin by a double adhesive ring. It is usually placed on the chest, back, thigh or abdomen of the neonate. The shaved-chest and forearm is a good location in adults. The specific details of design, calibration and its application have been described. (8)

Sensor Performance in Humans

The gestational age, weight and the diseases of the infants studied were not significantly different from those of the other two evaluations. The sensor, connected to a Roche model 6300 monitor, was prepared and calibrated according to the manufacturer's instructions. It was attached to the skin by a duble adhesive ring. The usual position for attachment was the subclavicular part of the thorax, the abdomen or the thighs. After application to the skin, 30 minutes were allowed to stabilize the sensor. The skin seal was tested by blowing 100 percent oxygen over the sensor. It was removed from the skin every four to six hours, recalibrated and applied to a new location. The electrode temperature was maintained at 44°C to cause hyperemia of the skin and "arterialization" of the capillary bed.

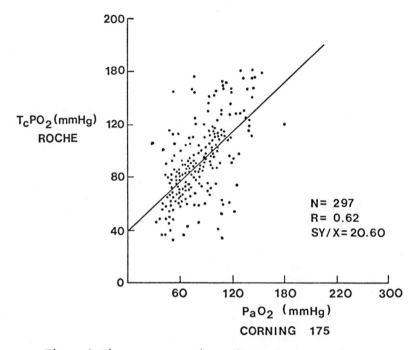

Fig. 8 Linear regression of the Roche Tc PO2 sensor and the PaO2 as measured with a laboratory analyzer. The correlation coefficient is 0.62 from 297
comparisons.

Transcutaneous measurements (TcPO2) were compared to PaO2
measured by a laboratory analyzer. The blood for PaO2 measurements
was sampled from a catheter introduced through the umbilical artery
into the aorta. A total of 297 paired values were obtained from
nine neonates. Linear regression analysis of simultaneous TcPO2
and corresponding laboratory values was done on the data pairs from
each patient and a correlation coefficient was computed (Table 2).
Figure 8 is a linear regression of the 297 data pairs; the correlation coefficient is 0.62.

In a similar evaluation, the Radiometer model TCM 1 monitor
and cutaneous electrode were applied to 13 neonates.

Forty-one data pairs were obtained and a regression analysis
performed. The correlation coefficient was 0.74.

DISCUSSION

The PO2 catheter sensor appears to be non-toxic, non-thrombogenic, safe and reasonably accurate. An extensive trial in humans is now appropriate. The component parts such as the membrane are attached securely enough to remain in place and intact during patient use. The rapid response time and accuracy in both arterial and venous (in one subject) circulation should permit detection of small intermittent shunts and arterio-venous differences. If the concentration and type of hemoglobin and the oxygen consumption are known, it makes possible continuous measurement of cardiac output by the Fick principle. A better approximation of cardiac output would be obtained if the PO2 sensor was used in combination with an accurate SO2 measuring device.

The oximeter catheter system detects changes in SaO2 rapidly and it is a better measure of oxygen content than PaO2 in the steep portion of the oxygen dissociation curve. This system cannot replace the measurement of PaO2 which might be greater than 100 mmHg when the SaO2 is above 92 percent. However, another study suggest that a PaO2 measurement is not necessary until the SaO2 is above 96 percent. (10) The accuracy of the catheter oximeter and the unique calibration system which enables it to be used immediately upon insertion, makes it a useful and desirable tool for intensive care monitoring and as a guide to administer oxygen during cardiopulmonary resuscitation.

Within the limitations described, both the PO2 and SO2 catheter sensors will facilitate prompt adjustment of (FiO2) and respiratory support devices. This will help to eliminate an unnecessarily high FiO2. Improved control of PaO2 in newborn infants will contribute to prevention of bronchopulmonary dysplasia and retrolental fibroplasia. Fewer blood samples and transfusions will be required in the neonate. This should decrease the number of patients receiving mismatched blood, hepatitis and cytomegalic inclusion virus disease.

The PO2 and SO2 catheter systems are designed with a central lumen. This eliminates the need for side holes, makes blood pressure monitoring more reliable and reduces the possibility of a blood clot obstructing the lumen, a frequent problem with side hole devices. (11) The simplicity of the design of both catheters is of commercial importance.

At least two types of TcPO2 sensors do not accurately represent PaO2, and could not be expected to measure more than trend changes. This technique certainly will not replace laboratory PaO2 measurements and as such, cannot be considered an adequate substitute for intravascular catheters.

In view of a continuing need for intravascular catheters, blood gas sensors which can be incorporated into them without interferring with the usual catheter function are of obvious importance. Catheters with this capability will undoubtedly help to improve patient care. Knowing the mixed venous SO2 or PO2 can alert the medical staff to significant changes in the cardiac output. (12) Airway occulsion, tubing disconnected from respiratory support devices or accidental changes in FiO2 can be immediately recognized. Nurses can use continously recorded data to determine when to suction endotracheal tubes. From this information they can also learn how frequently these procedures cause the patients blood oxygen level to become dangerously low. It appears that it might be possible to servo-regulate FiO2 using PaO2 and SaO2 measurements.

SUMMARY

PO2 and SaO2 measurements performed by the catheter sensors described in this report are accurate and provide useful data that can be computer processed. They can be used to monitor a patient's oxygenation and could be useful to servo-regulate the FiO2. The TcPO2 sensors do not accurately reflect PaO2 and cannot be recommended for patient monitoring when accuracy and reliability are of the utmost importance.

Table 1

Data Analysis of PO2 Catheter Sensor

Case	Number of Determinations	Correlation Coefficient R	Hours of use	Artery or Vein A or V	PaO2-mmHg range
Ma	10	0.88	10	V	16-39
Jo	7	0.86	20	A	117-193
Fr	37	0.86	51	A	44-290
Gr	16	0.79	30	A	83-196
Ea	16	0.79	38	A	58-97
Bl	13	0.71	81	A	68-153
Kr	18	0.66	30	A	69-153

Table 2

Data Analysis of Roche Cutaneous PO2 Sensor

Case	Number of Determinations	Correlation Coefficient R	PaO2 – mmHg range
So	13	0.87	60–111
Se	6	0.77	56–157
Je	30	0.70	60–135
Ri	80	0.63	33–200
Re	76	0.61	46–127
Ma	24	0.61	57–180
Co	8	0.61	47–97
Pi	38	0.53	44–147
At	22	0.45	44–105
Pooled data	297	0.62	33–200

REFERENCES

1. BROWN, EG, LIU, CC, MCDONNELL, FE, ET AL: Experimental Medicine and Biology. Bruley, DF and Bicher, HI, eds., New York: Plenum Press, pp 1103, 1973

2. CLARK, LC, JR: Monitor and control of blood and tissue oxygen tensions. Trans Am Soc Artif Int Organs 2:41, 1956

3. MCDONNELL, FE, BROWN, EG: Unpublished data

4. WOLFSON, SK, JR, LIU, CC, MCDONNELL, FE, BROWN, EG: Continuous arterial oxygen monitoring in the awake baboon (papio cynocephalus-anubis) using a unique oxygen electrode. Dig of the 11th Intl Conf in Med Biol Engrg, Ottawa Canada, p 62, 1976

5. KUSSEROW, B, LANOW, R, NICHOLS, J: Obervations concerning prosthesis - induced thromboembolic phenomena made with an in vivo embolus test system. Trans Am Soc Artif Orgnas 16:58, 1970

6. POLANYI, ML, HEHIR, RM: In vivo oximeter with fast dynamic respose. Rev Scient Instrum 33:1050, 1962

7. ENSON, Y, JAMESON, AG, COURNAND, A: Intracardiac oximetry in congenital heart disease. Circulation 29:499, 1964

8. EBERHARD, P, MINDT, W, JANN, F, HAMMACHER, K: Continuous PO_2 monitoring in the neonate by skin electrodes. Biomed Engrg 13:436, 1975

9. HUCH, R, HUCH, A, LUBBERS, DW: Transcutaneous measurement of blood PO_2 ($TcPO_2$) - method and application in perinatal medicine. J Perinat Med 1:183, 1973

10. WILKINSON, AR, PHIBBS, RH, GREGORY, GA: Continuous measurement of oxygen saturation in sick newborn infants. J. Pediatr 93: 1016, 1978

11. GODDARD, P, KEITH, I, MARCOVITCH, H, ROBERTON, HRC, ROLFE, P, SCOPES, JW: Use of a continuously recording intravascular oxygen electrode in the newborn. Arch Dis Child 49:853, 1974

12. HUGENHOLTZ, PG, KRAUSS, AH, VERDOUW, PD, NAUTA, J: Clinical experience with on-line fiber optic oximetry: Oxygen Measurements in Biology and Medicine. Ed. Payne, JP, Hill, DW. Butterworths, London, 1975, 383

COMPUTER ASSISTED EVALUATION AND COMPUTATION OF PULMONARY FUNCTION IN THE CRITICALLY ILL NEONATE

Jacob G. Schwartz, M.S., Andrew M. Trattner,
Thomas H. Shaffer, Ph.D., and William W. Fox, M.D.

The Division of Neonatology,
The Children's Hospital of Philadelphia

The Department of Pediatrics
University of Pennsylvania
School of Medicine

The Department of Physiology
Temple University
Philadelphia, Pa.

At the Infant Intensive Care Unit of The Children's Hospital of Philadelphia, approximately 500 newborn infants are admitted each year with various forms of respiratory disease, such as Respiratory Distress Syndrome, Meconium Aspiration Syndrome, and Neonatal Apnea. In these situations, when oxygen and carbon dioxide gas exchange is significantly compromised, it becomes necessary to apply respiratory support such as mechanical ventilation, continuous positive airway pressure (CPAP) and/or increased inspired oxygen concentration (FiO2) to maintain adequate vital signs.

One of the ways a clinician may assess the severity of disease and progress of a patient toward recovery is to perform pulmonary function tests. Examples are dynamic lung compliance, inspiratory and expiratory lung resistance, tidal volume, tidal esophageal pressure, minute ventilation and work of breathing. A common technique for obtaining these parameters is to measure esophageal pressure, tracheal air flow and volume. Esophageal pressure is measured using a miniature esophageal balloon connected to a polyvinylchloride catheter passed into the lower third of the esophagus. This pressure is then transmitted to a pressure transducer where it

121

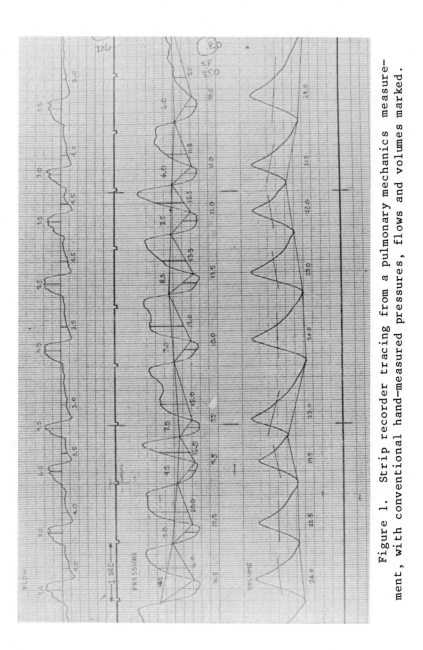

Figure 1. Strip recorder tracing from a pulmonary mechanics measurement, with conventional hand-measured pressures, flows and volumes marked.

is converted to an electrical signal. Tracheal air flow is measured using a pneumotachograph and flow transducer. Volume changes
during breathing are measured form electronic integration of air
flow, and all three signals are transmitted to a strip recording
device. With these techniques, signals can be obtained from infants weighing as little as one pound.

Using the strip output, an investigator must measure, by pen-
cil and straightedge, the desired parameters from up to ten consecutive uniform breaths. In order to measure all the respiratory
values from the strip chart, as shown in Figure 1, an hour or more
may be required, assuring careful calculation.   Other measuring
systems may display breathing loops or yield breath-to-breath results of one or two parameters, however, none evaluates all the values desired.   To alleviate this problem, a system of computer
software has been generated which samples the incoming signals from
the infant and automatically calculates pulmonary function parameters.

As shown in the schematic in Figure 2, the esophageal balloon,
B, a pneumotach Pn, and their respective pressure and flow transducers in PT and FT, are connected to the recorder which is represented by the dotted lines. Pressure and flow signals are amplified by amplifiers, PA and FA, and flow is integrated to volume by
the integrator, I. In addition, a cable, carrying the pressure and

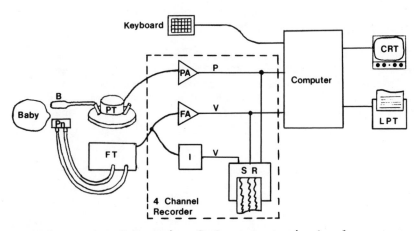

Figure 2.   Schematic of the computerized pulmonary
function testing system. P = pressure; V̇ = flow; V =
volume; SR = strip recorder.

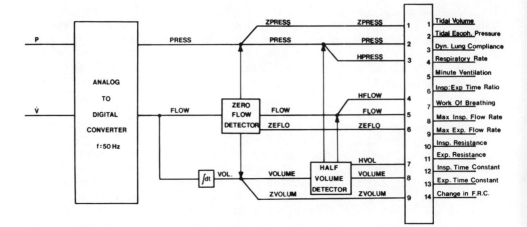

Figure 3.   Simplified schematic of the computer's
function in pulmonary mechanics computation.   P = pressure; V̇ = flow.

flow signals, connects the recorder to a computer. This is a Hewlett Packard 2100S minicomputer, which is currently the central processor of an HP 5600A Patient Monitoring System presently used for
measuring, storing and displaying vital signs and other important
information such as hemodynamic and blood gas data.

A handheld keyboard, located near the patient's bed, is used
to control the computer's function. The viewing screen (CRT) over
the bed shows information and a nearby printer (LPT) provides a
permanent record of viewing screen output.

The computer is signalled for respiratory measurements and the
screen displays "Press Go To Begin Sampling." The pneumotach is attached to the patient either by facemask or directly to his endotracheal tube, and the esophageal balloon is passed nasally. The
recorder is turned on and pressure, flow and volume are monitored
on the recorder paper until uniform breathing is achieved, and the
"GO" is pressed.

Figure 3 shows a simplified view of the computer's function.
The two signals are measured at a rate of 50 samples per second by
an analog to digital converter. The flow data is mathematically
integrated into volume, thus generating the three basic values;
pressure, flow and volume, from which all desired parameters are
calculated. Two other steps are performed by the computer to allow
it to generate the outputs. First, points of zero flow are detected and stored as a separate entity, called ZEFLO, and along with

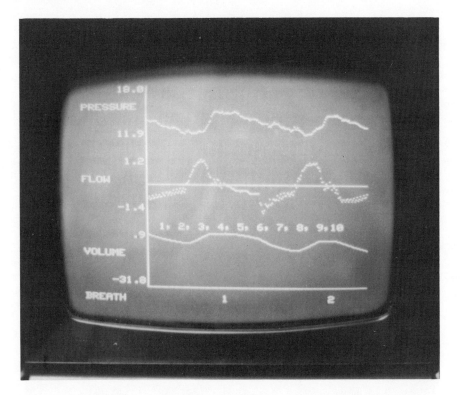

Figure 4. Pressure, flow and volume time plots shown on the viewing screen for the first two measured breaths. Numerical values along the vertical axis represent minimum and maximum limits of each signal. The number of the breaths chosen for calculation (in this case 1 through 10) appear on the screen between flow and volume traces.

it, the pressure and volume points occurring at no flow, called ZPRESS and ZVOLUM, respectively, are created. These are used in the calculation of dynamic lung compliance, which is the stiffness of the lung. Lung compliance is the ratio of the tidal volume and tidal esophageal pressure. Secondly, points of volume located at half the height of the tidal volume are found and labelled HVOL, and along with these, the pressure and flow points which occurred at half-volume are separated out to create HPRESS and HFLOW. These are utilized in the computation of inspiratory and expiratory resistance, which is the relationship between flow and pressure at points of equal inspiratory and expiratory volume. It is from these nine parameters that the 14 values, shown as outputs at the right of Figure 3, are calculated. This entire process requires about 30 seconds for completion.

The first display on the viewing screen following this sampling period is the time plots of the pressure, flow and volume signals, with the individual breaths automatically numbered. The physical limit of the screen only allows four of the 20 seconds of data to be displayed at one time, therefore we have made it possible to scan horizontally across the three signals, showing a four second window of the 20 seconds of information. The user, while scanning, decides which breaths are to be considered for calculation.

Figure 4 shows an example of the first four seconds of breathing of a 20 second measurement, showing the initial two breaths, numbered below. This infant weighed 1.5 kg., had hyaline membrane disease and had been spontaneously breathing on CPAP for three days. Pressure is shown in cmH2O, flow in L/min and volume in ml. By scanning once to the right, the next breath, number 3, would come into view, and further scanning would show all the remaining breaths. As one inspects the signals, the breaths from which results are desired can be chosen by number. The choices are entered by the keyboard and also shown on the viewing screen to double check.

An additional keyboard signal produces the results for the chosen breaths. This occurs immediately, since the computer has actually calculated all of the information previous to the choice of breaths being displayed on the screen. Results appear in four sections, with tidal volume, tidal esophageal pressure, dynamic lung compliance and lung compliance per kg. in the first section. The average values are also shown for the chosen breaths. Further results like lung resistances, flow rates, time constants, work of breathing and the others mentioned in the list of outputs (Fig. 3) appear in the other screen displays.

The user may return to calculate values from other combinations of breaths from the same 20 seconds of data if necessary. Finally, a printed report of the pulmonary function data for all the breaths in the measuring period is generated on the printer for permanent records. This printout duplicates the four sections of results which appear serially on the viewing screen. The strip recorder output from the same time period also serves as a reference for later investigation of patient results.

We have hand calculated pulmonary functions in a small series of patients by the conventional technique, and found results to be extremely close to computer-generated values. Because the computer resolution is four times greater than that of the strip recorder, our statistical correlations tended to actually assess the accuracy of the technician's hand, rather than the automated technique. Nevertheless, the correlaiton coefficients for 65 breaths were never below $r = 0.90$ for any parameter, with only one (expiratory resistance) below $r = 0.94$.

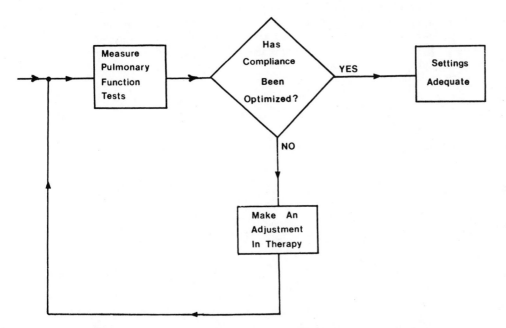

Figure 5.   An example of a respiratory   management
scheme   for   maintaining   a   patient at optimum dynamic
lung compliance.

The ability to obtain almost   instant   results   for   pulmonary
function   parameters may be more efficacious in determining whether
current respiratory support is adequate for a patient,   or   whether
modifications in ventilator settings may be beneficial or detrimen-
tal.   For instance, a clinician may wish to maintain a   patient   on
respiratory therapy   which   is optimized for the patient's dynamic
lung compliance.   In this case, he may follow a scheme as   the   one
shown   in   Figure   5.   After measuring pulmonary function tests and
evaluating compliance, he can adjust ventilator, CPAP or   inspired
oxygen   concentration, remeasure PFT's and repeat the process until
the compliance has   been   optimized.   Different   portions   of   the
20-second   measuring period may be viewed, and the effects of short
term breathing events may be assessed quickly and accurately.   For
example,   the computer allows calculation of pulmonary function values before and after "sigh" breaths, apnea, or the application   of
CPAP.

For research purposes, investigators may now compile results with much increased speed and reliability. In our own department in the past few years, we have hand-calculated results in hundred of patients to assess clinical care in applications such as the effect of prone positioning on pulmonary function and arterial oxygenation, physiologic effects of chest physiotherapy and suctioning, and pulmonary function changes with crying. Overall, the automation of this intricate and timetaking procedure is a significant step forward in improved patient care and management in the intensive care neonate.

COMPUTERIZED MONITORING OF LUNG AND HEART FUNCTION IN THE OPERATING

ROOM: A DESCRIPTION OF "METHODOLOGICAL PLATFORM".

O. Prakash, M.D.*,Senior Anaesthestist
B. Jonson, Ph.D.**, Clinical Physiologist
S.H. Meij, M.Sc.*, Computer Engineer
S.G. van der Borden, B.Sc.*, Research Assistant
P.G. Hugenholtz, M.D.*, Professor of Cardiology

* From the Thoraxcenter,
  Erasmus University and University Hospital
  Rotterdam, The Netherlands

** From the Department of Clinical Physiology,
   University of Lund,
   Lund, Sweden

Reprint request to: O. Prakash, M.D., Thoraxcenter,
                      Erasmus University
                      Postbus 1738, 3000 DR Rotterdam,
                      The Netherlands

3.A: Introduction

In the operating theatre, it has only recently become common-
place to monitor the elementary functions pertaining to the heart,
with the electrocardiogram, the arterial blood pressure, and the
pulse as the essential guidelines.

However, measurement of pulmonary function may be as vital as
that of cardiac function, not only to enhance the safety of the patient, but also to enable ventilation with controlled gas exchange
to be achieved. If this is done, the result will be maintenance of
the all-important carbon dioxide tension within normal limits. If,
however, the pulmonary function is not monitored, vital disturbances in ventilation or gas exchange (which may give more rapid or
early, accurate information concerning a developing crisis) may be
missed.

In addition to allowing accurate control of gas exchange and
stabilization of desired arterial carbon dioxide tensions, monitoring of pulmonary function has a number of other important applications. For example, by accurately predicting the future respiraory
performance of a sick patient, it helps in the correct timing of
the decision to replace controlled by spontaneous respiration as
the patient's condition improves. Moreover, establishing by monitoring, the physiological status of the patient also enables controlled deviations from the normal to be undertaken, as with progressive hypothermia in small infants undergoing cardiac surgery.
Adequate monitoring also permits study of the effects of various
physiological "insults" sometimes imposed on the human body; for
example, the effects of anaesthetic drugs, of transfusions, of surgery or accidental trauma. Postoperative threatening complications
can be recognized early and the effect of treatment against them
can be studied.

An outline of the results of having inadequate monitoring facilities

The anaesthetist must be constantly aware of the depth of anaesthesia being achieved while inhalation anaesthetic agents are
being administered. Variations in anaesthetic depth may lead to
serious disturbances in the patient's tidal and minute volume ventilation (if breathing spontaneously) or in blood and airway gas
composition (whether ventilation is spontaneous or
artificially-maintained). This in turn can lead to critical disturbances in the patient's cardiovascular status.

Thus, a knowledge of the patient's ventilatory performance and
airway gas composition must be regarded as fundamentally important
to the anaesthetist. However, experience shows that in the vast
majority of anaesthetics administered, patient-monitoring is restricted to a few simple cardiovascular and pulmonary parameters
like pulse, blood-pressure, capillary bed color and perhaps the
electrocardiogram, which will reveal only the end-results of ventilatory disturbances.

In 1976, a report was published in the Journal of the American
Medical Association (J.A.M.A.) entitled "Unexpected cardiac arrest
during anaesthesia and surgery".(1) In this report, 41 cases of
cardiac arrest which occurred during surgery, comprising 30 adults
and 11 children, were classified according to the most likely primary causative mechanism. Cardiac arrest was thought to have resulted from anaesthetic or ventilatory mismanagement in two-thirds
of the cases. In 19 of the patients, hypoxia was listed as the
cause of the arrest, and it appeared to be primarily due to hypoventilation. In 9 of the patients, the arrest was attributed to
anaesthetic mismanagement, while in a further 9 patients, severe
bradycardia which might have been secondary to other events, occurred. Four others had hypotension associated with hypovolaemia.

Of the total of 41 patients, 13 were dark-skinned, representing 31.7% of the group studied. This is interesting, since many anaesthetists depend only upon the repetitive observation of skin color as an index of adequate oxygenation and adequate peripheral perfusion. As a result of such a limited monitoring regimen, non-white patients may be at significantly greater risk than their white counterparts. This report subsequently came to the attenion of a major liability insurance company, a development spurred on by the fact that an earlier report(2) had already found an incidence of cardiac arrest in the operating area of 1 case in every 3400 operations.

In his recent article entitled "Monitoring in the operating room: Current techniques and future requirements", Hilberman (1977)(3) reviews the J.A.M.A. report and the subsequent lawsuit, and agrees that ventilatory problems appear to be a major source of intra-operative morbidity and mortality.

Obviously monitoring of ventilation could prevent most accidents of ventilatory origin. Furthermore, some circulatory problems with a fall in cardiac output are detected earlier from monitoring of gas exchange than from E.C.G. and intravascular pressure.

3.C: Modern requirements for the practice of para-operative pulmonary control

Hilberman's assessment of the present situation calls not only for equipment which will efficiently monitor cardiovascular parameters, but also for apparatus which will administer and monitor ventilation and airway gas concentrations with equal efficiency. He states that, in principle, these are a fundamental necessity in modern anaesthetic practice. Unfortunately, it is not certain that such equipment will be easy to promote, once built. The problem is that in order for monitoring devised especially to find ready acceptance among the majority of potential users (mainly anaesthetists and intensive care doctors) the devices must be small, simple and convenient to use, and it must be trustworthy, accurate, and relatively inexpensive. Such requirements are not easy to combine into a practical piece of commercial apparatus. In practical terms, the requirements for control of modern anaesthetic pulmonary ventilation may be considered under three headings: the mechanical ventilation itself, the efficiency of gas exchange, and the stability of pulmonary mechanics.

Until the second World War, accurate knowledge of the composition of expired gas could be provided only by chemical analysis. With the Haldane apparatus it took some 20 minutes for the analysis of a sample to be performed by an accomplished operator. The method had an accuracy of 0.1 mo 1% in skilled hands. By 1950, methods

based upon various types of thermal conductivity changes, infra-red
absorption (carbon dioxide and nitrous oxide), paramagnetic suscep-
tibility (oxygen) and spectral emission (nitrogen) were in use(4).
These instruments mostly had a long response time, with the exception of Lilly's nitrogen meter(5). It was also recognized at that
time that mass spectrometry was a promising analytical method for
respiratory gases, since it was specific for gases of different mo-
lecular weights. In 1947, Siri(6) scanned the mass range of medical gases with a simple and compact instrument at a repetition rate
of 200 Hz, and showed that single peaks could be recorded with response times of little greater than one second. Hunter, Stacey and
Hitchcock (1949)(7) made an instrument with a sample transport system showing a quarter-second response time, and with three fixed
collectors aligned to receive nitrogen, oxygen and carbon dioxide
simultaneously. In 1950 a single channel conventional instrument
was modified by the addition of a short sampling system of response
time under 100 ms and used in a number of respiratory studies in
the University of Pennsylvania.(8)

Although mass spectrometry has seen increasing use in respiratory research work and clinical investigation, its use in the operating theater remains limited by serious drawbacks. The sheer size
and high capital cost of a mass spectrometer generally makes it impractical for use in the operating room; it is often severely affected by volatile anaesthetic agents and it is technically difficult for the small "clinical" mass spectrometer to distinguish
between nitrous oxide and carbon dioxide, since both have a
mass-number of approximately 44 ($N2O = N+N+O = 14+14+16=44$: $CO2$
$=C+O+O+ = 12+16+16=44$).

Based on the work of Peters and Hilberman(9), whose elaborate
I.C.U.- based computerized pulmonary function monitoring system was
reported in 1971; Osborn, in 1977(10), developed a more refined
system for respiratory monitoring in the intensive care unit. He
used a computer to handle the processing of all the data he acquired just as Peters and Hilberman did. Of more than 30 parameters measured, he came to the conclusion that the most useful one
in the mechanically ventilated patient was the end-tidal $pCO2$
(PetCO2). This was simply because it is a direct measurement of
whether the preset minute volume ventilation is adequately "clearing" pulmonary arterial blood of its carbon dioxide load. In the
hypoventilated patient, this "clearing" process becomes inadequate,
and the PetCO2 will rise; by contrast, hyperventilation leads to a
"washing-out" of carbon dioxide, and hence the PetCO2 falls. Thus,
any mal-adjustment, leak or faulty function in the ventilator is
likely to show up promptly as a change in PetCO2, which makes this
measurement an extremely helpful one in routine clinical practice.

3.B.(iii): Monitoring of Pulmonary Mechanics

Although it is over 50 years since Fleisch first described the pneumotachograph, it is only in recent years that this technique has become widely accepted. This delay has been caused by the many practical problems which exist with the use of Fleisch heads, especially the deposition of water from humid gases (which upsets the characteristics of the particular instrument) and the changing physical constitution of the gas mixture passing through it (which changes the pressure/flow relationship of the instrument). Even today, Fleisch's equipment has some limitations to its usefulness in routine clinical circumstances. However, the evaluation of lung function must be based upon measurements of parameters such as expiratory and inspiratory flow rates, airway pressures, and respiratory gas concentrations. In a lung function laboratory, studies can be made using with and complicated instruments, operated by specially-trained technicians and laboratory assistants. Conventional equipment of this sort, is bulky, generally involves the use of extra tubing and pumps, and frequently still relies on the use of non-electronic calibration systems.

3.D: The Instrumentation Employed in the "Methodological Platform"

3.D.(i) Introduction

The comprehensive monitoring system developed at the Thorax-centre in Rotterdam, allows accurate study and control of pulmonary and cardiovascular functions during anaesthesia and in the immediate post-operative period. It has been called "The Methodological Platform".

The discussion in the previous section attached an importance to pulmonary monitoring and ventilation considerably greater than that normally considered necessary by clinicians. It is perhaps not surprising, therefore, that the focal point of the Methodological Platform became a modern, efficient pulmonary control system. However, the information produced from this system did not constitute the only data which was collected from the patient for presentation to the clinician. In order to help with his overall patient management, data was also collected from oxygen analyzers, intra-vascular catheter/pressure transducer combinations, E.C.G. machines and temperature probes, and displayed for the clinician, as were the results the results for blood gas analysis. In many cases, measurement of cardiac output was performed.

A unified and consistant approach produces data which may be regarded as a basis upon which clinical decision-making can be based, and from which many different rational treatment strategies can be devised and pursued. Hence, it may truly be said to constitute a "Methodological Platform". The success or failure of such a concept depends not only on the quality, but also on the relevance of the data and on the ease with which they may be interpreted.

3.D.(ii): The Central Role of the Servo 900 B Ventilatior and its Monitors

Although there are a number of other important variables which are measured or derived, the variable of central importance in Critical Care patient management is considered to be the status of the blood acid-base biochemistry. Upon this rests, for example, the important decision when respiratory insufficiency necessitates controlled ventilaion or an alteration in ventilatory pattern; or conversely, when an improvement in status allows resumption of spontaneous respiration. Also resting upon it is the control of blood chemistry during any normal physiological excursions which may be imposed upon the patient - - for example, deep hypothermia.

3.D.(iii): The Ventilator

The central unit is the Servo 900 B ventilator, which is a sophisticated electronic servo-controlled machine alllowing great flexibility and precision of ventilatory control. It incorporates pheumotachograph transducers, specially designed to overcome most of their drawbacks, and to monitor both inspiratory and expiratory flow-rates. These transducers control the opening of the inspiratory and expiratory valves, so regulating accurately the inspired volume and flow patterns; and the expired flow pattern, another transducer monitors airway pressure. Thus, the machine gives the clinician direct control over a large number of ventilator parameters, viz:

i)   Airway pressure;

ii)   Respiratory frequency;

iii)   Inspiratory time;

iv)   Pause time;

v)   Maximum expiratory flow;

vi)   Expired minute volume.

The parameters ii, iii and iv are set directly on the machine, while parameters i and vi are monitored continously on analog meters. The ventilator also has facilities for Intermittent Mandatory Ventilation (I.M.V.) and for immediate spontaneous respiration; in the last mode of action, the patient is able to breathe by himself for an intermittent flow of fresh gas which has been humidified by the ventilator, while his expired gases still pass through the measuring part of the ventilator. There is also a comprehensive alarm system incorporated.

The Servo 900 B ventilator thus represents, in our view, an almost ideal and flexible means of maintaining proper lung function both in ventilated and spontaneously-breathing patients. It is small, quiet, and easy to set up and use. It uses electronic transducers instead of mechanical systems for the measurement of pressure and flow. Thus, these parameters are readily available for control, for analysis, and for display. Finally, it has a readily and comprehensive alarm system built into it.

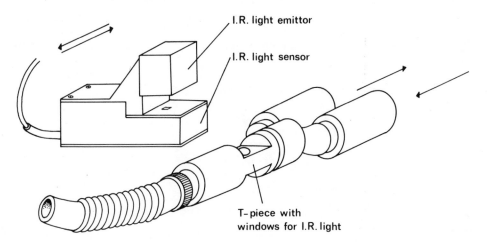

I.R. light emittor

I.R. light sensor

T-piece with windows for I.R. light

Fig. 1 The "Cuvette" constitutes a Y-piece connecting the expiratory and inspiratory lines with the patient. The cuvette is attached to the photometer so that one window in its wall faces an infrared emitter and the opposite window an infrared sensor.

3.D.(iv): The Carbon Dioxide Analyser

As has already been pointed out, the main function of the lungs is for respiratory gas exchange. Hence, it appears much more logical to control the gas tensions achieved by a given ventilatory regime than to simply control the ventilator settings themselves in some arbitrary "rule-of-thumb" fashion with, perhaps, occasional blood-gas analysis. Osborn (1977)(10) had discussed the results which demonstrated that in all normal clinical circumstances, the end-tidal carbon dioxide concentration (PetCO2) provides an accurate reflection of arterial carbon dioxide tension (PaCO2).

In the first place, the fractional expired carbon dioxide concentration (FeCO2) is measured through a highly selective, miniaturized infra-red analyzer unit. This analyzes the whole of the patient's gas flow, since it scans across the ventilatory Y-piece itself; this avoids all the problems and potentially dangerous complications which can arise from the withdrawal of gas samples from the main air-flow. The infra-red radiation passes across the Y-piece via sapphire glass windows, which are easy to keep clean. The radiation then passes via a narrow-band optical filter to a Gallium-Arsenide solid state detector(Fig 1). The moment-by-moment expired CO2 is immediately available as a continuous signal. In addition, by utilizing the timing pulses from the ventilator, the unit is able to display a breath-by-breath value for the end-tidal CO2.

Secondly, the 900 B ventilator yields an expiratory flow signal which, by integration with the instantaneous CO2 signal from the unit 930, will produce a signal representing moment-by-moment expired flow of CO2, and further integration a integration a value for expired volume of CO2, breath-by-breath or per minute. The tidal volume is divided into two parts, namely the initial "ineffective" one that did not reach alveoli and the "effective" one that took part in gas exchange and, hence contains CO2.

All in all the 930 provides information on:

i)   The end-tidal CO2 concentration;

ii)   The tidal CO2 production;

iii) The minute CO2 production;

iv)   The ineffective tidal volume;

v)   The effective tidal volume;

vi)   The effective minute volume.

It thus represents a major step forward in the field of respiratory gas monitoring in conjunction with mechanical ventilation.

3.D.(v):   The Lung Mechanics Calculator

The Lung Mechanics Unit 940 is another unit in the servo 900 B system. It will be remembered that in the Servo 900 B ventilator, the airway pressure is monitored by an electronic pressure transducer, while the gas flows into and out of the patient are measured by pneumotachograph heads. The electronic values of these parameters are immediately available on the output socket from the ventilator, together with timing signals derived from the Electronic

Unit. The Unit 940 connects directly to this output socket, from which it derives pressure, flow and time signals; on the basis of these, it calculates six different lung mechanics parameters. Incidentally, the raw signals from the Ventilator are also available for direct output to, for example, a paper chart recorder, oscilloscope or computer. This option was often used in these studies.

The parameters calculated by the Unit 940 are as follows:

i)   The peak inspiratory pressure;

ii)   The inspiratory pause pressure;

iii)   The end-inspiratory resistance;

iv)   The early-expiratory resistance;

v)   The pulmonary compliance;

vi)   The end-expiratory airway pressure.

The parameters are presented in pairs, the appropriate pair being selected by a switch on the front of the unit. The peak and pause pressures are values of the pressure signal sampled at the end of inspiration and at the end of the inspiratory pause plateau (Fig 2 & 3).

OPERATING THE SERVO 900 B SYSTEM

Because of the presence of the carbon dioxide analyzer, setting up the servo ventilator becomes very straightforward under normal clinical circumstances. A suitable respiratory rate is chosen and then the minute volume ventilation is adjusted by changing the tidal volume until a desired end-tidal CO_2 level appears on the unit 930. However, in the presence of severe lung disease, this approach may have limitations, since end-tidal CO_2 will not then approximate to arterial pCO_2 levels(11). The availability of extra derived data from the two units often allows detection of pulmonary malfunction so that fallacies can be avoided.

The minute volume ventilation necessary to maintain a "normal" $PaCO_2$ can be obtained from Nomograms 12-14; once it has been decided that the patient's ventilatory requirements appear to lie well outside his predicted values, the data from the CO_2 unit is used in a logical search for the underlying cause. THe CO_2 production may be chosen as a starting point. If this is higher than expected, the need for ventilation will have correspondingly increased. Further research should be directed towards factors influencing aerobic metabolic rate such as pain, anxiety, shivering, fever etc.

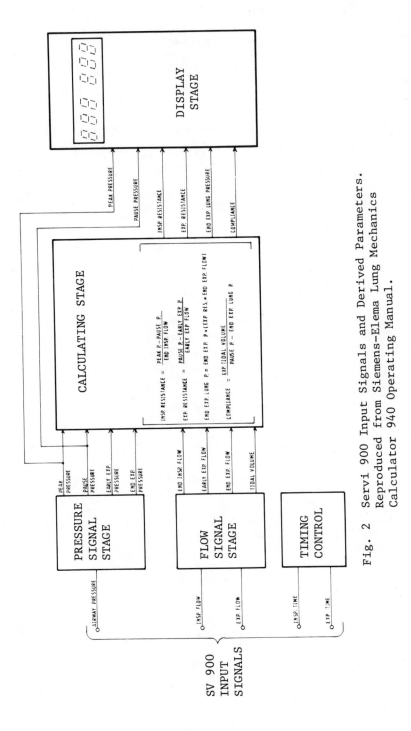

Fig. 2   Servi 900 Input Signals and Derived Parameters.
Reproduced from Siemens-Elema Lung Mechanics
Calculator 940 Operating Manual.

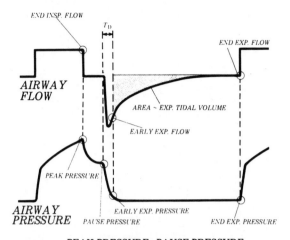

$$\text{INSP. RESISTANCE} = \frac{\text{PEAK PRESSURE} - \text{PAUSE PRESSURE}}{\text{END INSP. FLOW}}$$

$$\text{EXP. RESISTANCE} = \frac{\text{PAUSE PRESSURE} - \text{EARLY EXP. PRESSURE}}{\text{EARLY EXP. FLOW}}$$

END EXP. LUNG PRESSURE = END EXP. PRESSURE + (END EXP. FLOW x
EXP. RESISTANCE)

$$\text{COMPLIANCE} = \frac{\text{EXP. TIDAL VOLUME}}{\text{PAUSE PRESSURE} - \text{END EXP. LUNG PRESSURE}}$$

Fig. 3 Schematic pressure and flow curves, together with a summary of the method of derivation of
the various parameters mathematically. Reproduced from
Siemens-Elema Lung Mechanics Calculator 940 Operating
Manual.

In other cases, the alveolar ventilation, VA, is found to be
low. In the presence of normal total ventilation, VT, the physiological dead-space, VD, must therefore be increased; this can be
calculated from available data (see below). An increasing VD may
be caused by a high ineffective ventilation (absolute dead
space)(15); for example, an increased compressed volume in the
tubing or humidifier, caused by high airway pressures. This problem is of course of much greater importance in children, especially
in infants(16). The lung mechanics calculator will give additional
information about lung compliance and resistance, which may be the
cause of an increased airway pressure.

If the reason for the dysfunction has still not been found,
increased intrapulmonary or alveolar dead space must be present.
As is well known, such dead space can be caused by several factors:
Ventilation-perfusion mismatching, pulmonary embolism and venous
admixture (i.e. right to left shunt) may all contribute.

Various forms of dead space can be derived from the unit alone or in combination with arterial pCO2. Hence physiological dead space is calculated as:

$$V_{D \text{ physiol}} = V_T - \frac{V_{CO_2}}{Pa_{CO_2} \times K}$$

VT, VCO2 represent tidal volume and CO2 volume expired per breath and are obtained from the CO2 unit; K is a constant converting PaCO2 to fraction of CO2, i.e. expired CO2 concentration plotted during a breath against expired volume, on an oscilloscope.

The shape of the single breath curve gives information about ventilation/perfusion mismatching, since emptying of lung compartments with different ventilation-perfusion relationship yields a poorly-defined alveolar plateau. This type of pathological response is often found in obstructive lung diease, even when other relevant signs are missing. The CO2 analyzer therefore also offers the opportunity for study of these factors.

Non-sequential phenomena, such as venous admixture or synchronous emptying of lung compartments with different CO2 contents, yield a different pattern with a flat alveolar plateau that is lower than that corresponding to PaCO2. This pattern has also been observed in conjunction with lung embolism (non-perfusion of lung (compartments), and with variable right-left shunt(17). The usefulness and limitations of the single breath curve in detailed diagnosis of lung failure largely to be explored.

3.E: Other Instrumentation

3.E.(i):   Other Gas Analyzers

Where necessary, the inspired and mixed expired oxygen tensions, FiO2 and FeO2 were measured at the patient's airway connection (mouth or endotrachael tube) using a Servomex paramagnetic analyzer (Taylor-Sybron, England). With this analyzer used, an accuracy of +- 0.05% O2 can be achieved. A system in the analyzer line for drying the gases causes some mixing of the gases, and also introduces a time-delay in the display of the results. Thus only the mixed-expired oxygen concentration can be displayed.

After 1976, a Perkin-Elmer clinical Mass Spectrometer was used for the measurement of oxygen concentrations; with this instrument, nitrous oxide or carbon idoxide concentrations could also be measured simultaneously.

3.E.(ii): Cardiovascular Monitoring

Fluid-filled catheters in a systemic artery, in the right and
left atria, and often also in the pulmonary artery, allowed comprehensive monitoring of the various vascular pressures and cardiac
performance parameters.

The cardiac output could be measured directly using the Fick
principle. Blood samples were taken from the catheters for
blood-gas analysis, biochemical and hematological investigations.
The presence of the right and left heart catheters also allowed detection of left-to- right or right-to-left shunts.   The pulmonary
venous admixture (Qs/Qt) was calculated from the Equation:

where:

$\dfrac{Qs}{Qt}$ $\dfrac{Cc'O2 - CaO2}{Cc'O2 - CvO2}$ (2)

Cc'O2 = end-pulmonary capillary
             oxygen content

CaO2 = arterial oxygen content

CvO2 = mixed venous oxygen content

End-capillary oxygen saturation (Sc'O2) was calculated from
the "ideal" alveolar pressure (PAO2), the arterial carbon dioxide
tension (PaCO2) and the pHa, with the approach suggested by
Kelman(18). The "ideal" alveolar oxygen pressure was first calculated from the Equation:

$$PAO_2 = PiO_2 - \frac{PaCO_2}{R.Q.} \left[1 - FIO_2 \text{ x } (1 - R.Q.) \right] \qquad (3)$$

The oxygen content (vol%) was derived from the Equation:

$$C_{O_2} = \frac{SO_2}{100} \text{ x } (Hb \text{ x } 1.39) + (0.0031 \text{ x } pO_2) \qquad (4)$$

Oxygen saturation were also to be determined from a Hemoreflector
(American Optical Company), and in some instances continuously by
means of a Fibre- Optic Haemoreflector (Schwarzer Company, Munich).

From these data calculation of the right and left ventricular
stroke work and estimation of the perfusion resistances in the pulmonary and systemic circulations was carried out.   Left atrial
pressure was measured directly by inserting a catheter in the left
atrium after opening of the chest; the right atrial pressure was
taken to be equal to the central venous pressure.

Information regarding the quality of peripheral systemic perfusion was obtained by using a series of thermocouple temperature
probes (Ellab) positioned in the rectum or oesophagus (for core
temperture) and on the skin of the big toe. The skin probes were
kept covered by gauze wrappings, so as to minimize air-draught effects.

3.E.(iii):   Record Keeping

Manual record-keeping although essential is often unsatisfac-
tory.   Usually this is due to lack of the necessary time since attention is often withdrawn from the patient. Automatic
multi-channel recording machines are bulky and expensive. Although
central, fully computer-based, record-keeping has great attraction,
it requires a high initial outlay and is subject to intolerable
drawbacks in cases of system break-down. At the Thoraxcentre, at
the present time, a dual system based upon manual as well as automated record keeping is in use.

In the post-operative care unit in particular, the previously
described intensive care computer monitoring system -- the ICPM
system 17 -- allows record-keeping of physiological signals over 24
hours. The data inputs into the system include cardiac rhythm,
various intravascular pressures, and the signals from the Ventila-
tor/Lung Mechanics/Carbon Dioxide Computation System. All calculations (like cardiac output studies and trends) are performed on a
Digital Equipment Corporation PDP 9 computer, and became available
via a lineprinter or plotter at a later time. Current efforts
being made are to decentralize this form of bookkeeping to render
it in a condensed form under control of the nursing staff(20,21).

DISCUSSION

The introduction of multiple, minaturized, sophisticated monitoring devices opens up considerable possibilities. Numerical,
graphic or digital presentation of data relating to ventilation,
pulmonary mechanics, pulmonary gas exchange and cardiovascular performance allows the continuous monitoring of the main vital functions. Because of this, early detection and analysis of technical
or organic malfunction becomes possible. Several workers have, of
course, pointed out various applications of modern diagnostic techniques both during and after surgery, but the field has only been
explored to a very limited extent.

The present system has been built up in response to an increasingly felt need for continuous and detailed information concern-

ing cardiopulmonary status during and after open heart surgery.
Elaborate measurement systems may, however, convey problems of all
kinds: cabinets full of knobs for calibration procedures and settings; tubing, pumps, chart recorders, computers---space fully
taken up by equipment and personnel, leaving little room over for
anything except a caricature of a patient deserted in a sea of
technology! In contrast to this picture, the comprehensive use of
miniaturized and of electronic processing techniques has enabled
the present monitoring hardware to remain small, simple to set up
and use, and clear in its presentation of the important data.
Because of this, the welfare of the patient has remained the focus
of clinical attention.

In a unified and consistant approach all data must be combined
to provide a "Platform" upon which clinical decision-making can be
based and from which rational treatment strategies can be devised
and pursued. The success or failure of such a concept depends not
only on the quality of each signal but also on the relevance of the
data. Finally there is the factor of the display and on the ease
with which data obtained can be interpreted.

A brief summary of the various parameters which are monitored
routinely, and of the derivations of the processed data which is
then available, is shown schematically in Table 1.(a), (b), and
(c). The aim of the "Platform" is to permit successful monitoring
of all these parameters, with clear and concise data display; when
this aim has been realized in practice, then a truly quantitative
interface may be said to exist between the patient and his clinician.

A number of specialized clinical studies have been performed,
with a view to trying to establish just such a quantitative clinical interface using the "Methodological Platform". Studies have
been performed on early postoperative extubation(22), metabolic
changes during deep hypothermia(17), and hemodynamic and biochemical variables after induction of anesthesia with fentanyl/nitrous
oxide in patients undergoing coronary artery bypass surgery (23).
Some interesting observations have also been made on the occurrence
of various intra- and post-operative complications. As a result of
these studies, problems in cardiac surgery, cardiac anesthesia and
post-operative care have been reduced. Indeed, in some cases this
has turned out to represent an unexpectedly large improvement in
patient welfare. Moreover, the ability of the "Platform" to fur-
nish data which allows strategic manipulation of the patient's physiology has led to the discovery of a number of novel solutions to
well-known clinical problems.

Table I. A SUMMARY OF THE METHODOLOGICAL PLATFORM

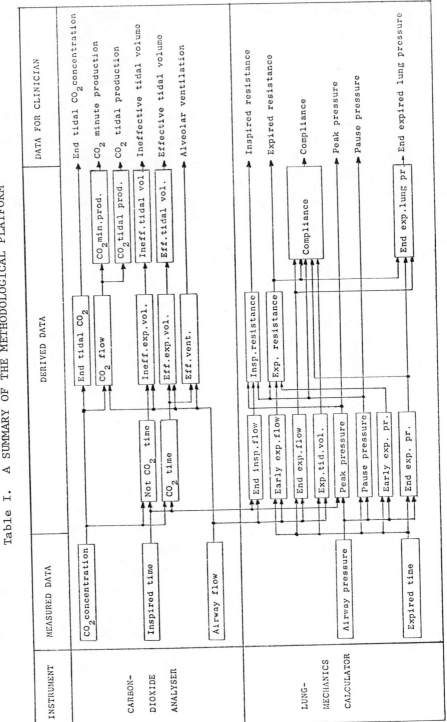

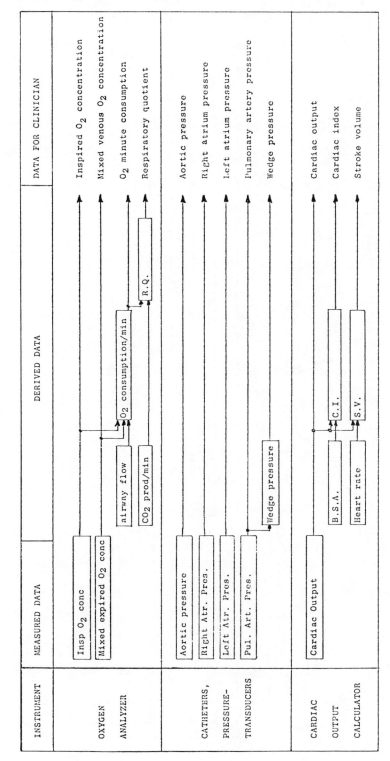

INSTRUMENT	MEASURED DATA	DERIVED DATA	DATA FOR CLINICIAN
OXYGEN ANALYZER	Insp O₂ conc Mixed expired O₂ conc airway flow CO2 prod/min	O₂ consumption/min R.Q.	Inspired O₂ concentration Mixed venous O₂ concentration O₂ minute consumption Respiratory quotient
CATHETERS, PRESSURE-TRANSDUCERS	Aortic pressure Right Atr. Pres. Left Atr. Pres. Pul. Art. Pres.	Wedge pressure	Aortic pressure Right atrium pressure Left atrium pressure Pulmonary artery pressure Wedge pressure
CARDIAC OUTPUT CALCULATOR	Cardiac Output	B.S.A. Heart rate C.I. S.V.	Cardiac output Cardiac index Stroke volume

Table I. (Continued)

INSTRUMENT	MEASURED DATA	DERIVED DATA	DATA FOR CLINICIAN
E.C.G.	E.C.G.		Electrocardiogram
MONITOR		Heart rate	Heart rate
TEMPERATURE	T_{core}		Core temperature
TRANSDUCERS	T_{toe}	$T_{peripheral}$	Peripheral temperature

REFERENCES

1. TAYLOR G, LARSON CP JR, PRESTWICH R. Unexpected cardiac arrest
      during anaesthesia and surgery. An environmental study
      J.A.M.A. 1976, 236,2758.

2. MC CLURE, J.N., SKARDASIS, G.M., BRUIN, J.M.: Cardiac arrest in
      the operating area. Amer Surgeon 1972, 38, 241.

3. HILBERMAN, M: Monitoring in the operating room: Current Tech-
      niques and future requirements. Med Instr 1977, 11, 283.

4. LILLY, J.C.: Physical methods of respiratory gas analysis Meth
      in Med Res 1950, 2, 133.

5. LILLY, J.C.: Mixing of gases within respiratory system with a
      new type of nitrogen meter, Am J. Phyiol 1950, 161, 342.

6. SIRI, W.: A mass spectroscope for analysis in the low mass range.
      Rev Sci Instrum 1947, 18, 540.

7. HUNTER, J.A., STACY, R.W., HITCHCOCK, F.A.: A mass spectrometer
      for continuous gas analysis. Rev Sci Instrum 1949, 20, 333.

8. COMROE, J.H. JR: The functions of the lung. Harvey Lec 1952, 48,
      110.

9. PETERS, R.M., HILBERMAN, M.: Respiratory insufficiency: Dia-
      gnosis and control of therapy. Surgery 1971, 70, 280.

10. OSBORN, J.J.: Cardiopulmonary monitoring in the respiratory
      intensive care unit. Med. Instr 1977, 11, 278.

11. FLETCHER, R., JONSON, B.: Skattning av arteriellt pCO2 genom
      matning av CO2 i expirationsluft. Svenska Lakaresall-
      skapets Riksstamma, p 84, Stockholm: Hygiea.

12. ENGSTROM, E.C., HERZOG, P.: Ventilation nomogram for practical
      use with the Engstrom respirator. Acta chir Scand 1959,
      Suppl 245, 37.

13. ENGSTROM, C.G., HERZOG, P., NORLANDER, O.P., SWENSSON, S.A.:
      Ventilation nomogram for the newborn and small children to
      be used with the Engstrom respirator. Acta anaesth
      Scand 1962, 6, 175.

14. RADFORD, E.P.: Ventilation standards for use in artifical re-
        spiration. J. Appl Physiol 1955, 7, 451. 15. BARTELS,
        F., SEVERINGHAUS, J.W., FORSTER, R.E., ETA AL: The
        respiratory dead space measured by single breath analysis
        of oxygen, carbon dioxide, nitrogen or helium. J clin
        Invest 1954, 33, 41.

15. OKMIAN, L.G.: Artifical ventilation by respirator for newborn
        infants during anaesthesia. Acta anaesth Scandinav
        1963, 7, 31.

16. PRAKASH, O., JONSON, B., MEIJ, S. et Al: Cardiorespiratory and
        metabolic effects of profound hypothermia. Crit Care
        Med 1978, 6, 340.

17. KELMAN, G.R.: Digital computer subroutine for the conversion of
        oxygen tension into saturation. J. Appl Physiol 1966, 21,
        1375.

18. HUGENHOLTZ, P.G., MILLER, A.C., KRAUSS, Z.H., HAGEMEIJER, F.:
        Automation in the management of data in the intensive care.
        Proc Eight Pfizer Int Symp in Edinborg 1973.

19. DEUTSCH, L.S., ENGELSE, W.A.H., ZEELENBERG. C., et al: The
        Unibed partient monitoring computer network. A new ap-
        proach for a new technology. Med Instr 1977, 11, 274.

20. ZEELENBURG, C., ENGELSE, W.A.H., DEUTSCH.L.S.: A hierachical
        patient monitoring computer network, Computers in Card-
        iology, 1977, 439. IEEE Computer Society, Long Beach,
        California.

21. PRAKASH, O., JONSON, B., MEIJ, S.: Criteria for early extuba-
        tion after intracardiac surgery in adults. Anesth Analg
        1977, 57, 703.

22. PRAKASH, O., VERDOUW P.D., DE JONG, J.W., et al: Hemodynamic
        and biochemical variables after induction of anaesthesia
        with fentanyl/nitrous oxide in patients undergoing cor-
        onary artery by-pass surgery.

OFF-LINE INTERMITTENT COMPUTERIZED CARDIOPULMONARY EVALUATION OF

SURGICAL PATIENTS

M. Klain

University of Pittsburgh
School of Medicine
Health Center Hospitals
1060-C Scaife Hall
Pittsburgh, PA 15261

INTRODUCTION

In the last few years there is an increasing trend to monitor in critically ill patients not only the blood pressure, pulse and ECG, but to evaluate the functional status of cardiorespiratory system; that means the cardiac performance and oxygen transfer and utilization. Using large computer systems we can generate a sizable amount of data especially with continuous monitoring, but this method is usually limited to big institutions and requires specialized computer personnel. Therefore we decided to implement a simple computer system in a so-called Automated Physiologic Profile Analysis (PPA). This method, introduced by J. D. Cohn and L. DelGuercio, was successfully used at Montefiore Hospital in Pittsburgh for the past four years.

A bedside data acquisition cart is used for measuring the arterial, venous and pulmonary artery pressures and performing the cardiac output determinations. We are using a modified Datascope cart which can be brought easily into any ICU room or to the operating room and contains a Datascope pressure monitoring unit and Waters dye dilution and thermal dilution unit. Only an arterial line and central venous line and/or Swan/Ganz catheter are needed.

Figure 1 - Physiologic Profile Analysis

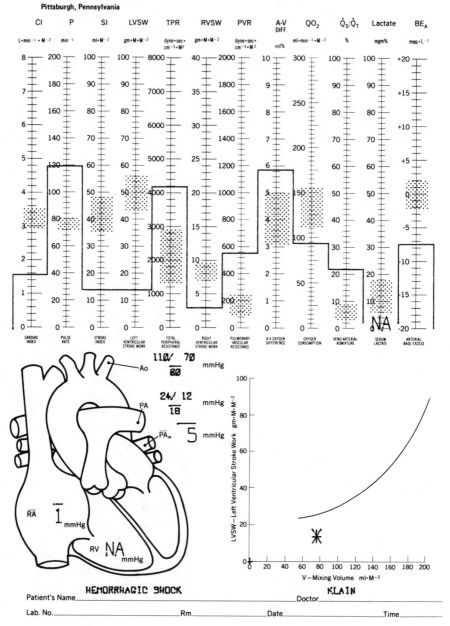

HEMORRHAGIC SHOCK

Patient's Name_____ Doctor KLAIN _____

Lab. No._____ Rm._____ Date_____ Time_____

The results of these measurements and of the blood gas analysis are brought to the closely located laboratory and put through
a small computer system. This consists of a programmable desk top
computer Wang "600" with tape memory, input/output writer and a
graphic digitizer. The dye dilution curve is traced on the graphic
digitizer which transfers the data to the computer. The computer
in turn calculates the cardiopulmonary parameters and prints all
the results on an alfa/numerical X-Y plotter.

The generated graphic plot uses a format similar to the biochemical profile. In the upper part of the graph, 12 parameters
are shown, seven of them related to the heart function and five to
the oxygen transport. They are: Cardiac index, pulse rate, stroke
index, left ventricular stroke work, total peripheral resistance,
right ventricular stroke work, pulmonary vascular resistance, arterial-venous oxygen difference, oxygen consumption, veno-arterial
admixture, arterial base excess, and finally, the serum lactate as
an expression of the tissue perfusion.

In the lower part of the profile graph are the numerical values of pressures measured in various heart compartments, the P-50
value which defines the shift in oxygen dissociation curve and the
heart function plot. At the same time the computer prints out on
the input/output writer a formal letter report and lists all the
parameters which differ from normal values.

The whole procedure of the Physiologic Profile measurements
takes about 15 minutes and the results from the computer may be delivered in another 15 minutes. Since we have studied the first patient in February 1975, we have performed more than 2,,500 automated physiologic profiles at the rates of about 16 per week but sometimes as many as 10 per day. We are providing this service now on
a 24 hours a day, seven days a week basis, mostly in the intensive
care units but also as a part of a pre-operative evaluation of the
critically ill patients or in the operating room if needed.

USE OF PPA

In our experience the value of Physiologic Profile lies in 3
aspects:

1. Helps to establish the diagnosis

2. Helps to establish proper treatment

3. Can be used for monitoring of critically ill patients in
   the operating room.

Initial or pre-operative diagnosis and identification of high risk patients is the first use of the Profile. Patients admitted in shock usually have low stroke index, compensated only partially by tachycardia and have signs of decreased oxygen delivery and acidosis.

After proper blood and fluid replacement and surgery, if needed, the hemodynamic values improve usually to, and above, normal range. It should be noted that in any patients despite a significant increase of cardiac index there is no major change in central venous pressure. This confirms that central venous pressure is not the best parameter for monitoring or estimating the blood volume and/or cardiac index. It should also be noted that these patients remain often acidotic despite the improved cardiac index. It can be considered a favorable compensatory mechanism for releasing oxygen in the periphery.

The profile can also help in differential diagnosis. The patient admitted with the diagnosis of septic shock can have a very high or a very low cardiac output. In our experience, most of the critically ill patients have complex alterations in their cardiopulmonary function which are sometimes difficult to detect by standard clinical and laboratory evaluation. Especially in compounded clinical pictures, a sequential PPA study allows to differentiate the individual participating factors and can quantitate the improvement achieved by the therapy. The changes of cardiopulmonary function in individual patients can not be predicted from clinical signs. In our experience none of the indirect methods estimating cardiac function from central venous oxygen, urinary output, or from arterial pressure, is a reliable indication of the true cardiac function.

The second use of Physiologic Profile is as a valuable tool to establish and monitor proper treatment. Especially in shock, the contributing factors of acidosis, cardiac failure, and peripheral resistance changes can be unmasked by sequential studies and the treatment by cardiotonic vasodilating drugs can be titrated. Not only can favorable effects normalizing the cardiac output and peripheral resistance and improving the oxygen delivery be demonstrated, but the optimalization of the treatment can be achieved. Patients admitted to the ICU with respiratory insufficiency and put on controlled ventilation with incremental levels of PEEP can be easily followed by Physiologic Profile determinations to establish the best PEEP which gives a significant improvement in pulmonary shunting without impairing the circulation.

Last but not least, the monitoring of critically ill patients in the operating room is an important part of the PPA use. The response of the patient to the operative trauma can be evaluated and proper treatment can be initiated. This applies especially to the use of blood and plasma expanders.

CONCLUSION

Our experience in more than 2,500 studies in the past four years has shown the following advantages of the Physiologic Profile Analysis:

(1) The graphic display -- most of the parameters have been used for monitoring and evaluation of the patients before, but the simple, understandable display of the profile is much easier to evaluate than the regular printed reports. The graphic displays have been accepted bery well by all the attending physicians.

(2) The PPA induces a different way of thinking. It gives us a tool to quantify the data as opposed to clinical impressions. The evaluation and monitoring of surgical patients in the past was based more on biochemical laboratory values like SMA 12 and on EKG monitoring and did not pay enough attention to the functional status of the circuulatory and respiratory system. The profile allows more emphasis to be put on the function rather than on the electrical activities or laboratory results.

(3) It also helps to evaluate the individual patient who often differs from the text book picture. The cardiac output, for example, can be decreased with the increasing level of PEEP, but in many of our patients the cardiac output did increase instead of decrease with an incremental level of PEEP. In some patients,also, the AV oxygen difference increased with increasing the cardiac index which is the opposite to the common belief.

(4) A better understanding of the functional status of the cardio-respiratory system might help us to better define our monitoring needs and help us to understand the patho-physiologic changes which occur. It might help us to re-define the disease classification in some cases. The unequal changes occuring in the stroke index, cardiac index and peripheral resistance are one example.

(5) Last, the prediction of operative mortality by evaluation of cardiac reserve is a valuable part of the PPA. The differences in so-called "mixing volume" might be the beginning of better understanding the cardiac reserve. Whether it is expressed in terms of mixing volume, cardiac contractility, residual volume or the mean transit time, it helps us to identify the high risk patients who require intensive treatment to decrease the surgical risk and the mortality as well.

TABLE 1   PHYSIOLOGIC PROFILE PARAMETERS

Parameter & Units	Normal Range	Derivation
Cardiac Index	2.9 - 3.5	$CI = \dfrac{\text{Cardiac Output}}{\text{Body Surface Area}}$

CI
$L \cdot min^{-1} \cdot M^{-2}$
Prime determinant of hemodynamic function. Elevated in septic shock.
Decreased in hypovalemia, pulmonary thromboembolism, cardiogenic shock. Derived
from green dye, thermal or Fick method.

Pulse	68 - 80	P = Heart rate derived from electrocardiogram

P
min^{-1}
Heart rate and rhythm noted during automated physiologic profile analysis.
Heart rate is utilized in derived data analysis.

Stroke Index	35 - 48	$SI = \dfrac{CI}{\text{Heart Rate}}$

SI
$ml \cdot M^{-2}$
Quantitative measure of ejected ventricular volume. Elevated in septic shock,
bradycardia, hypervolemia, hypertension. Decreased in cardiogenic shock,
hemorrhagic shock. Fixed stroke index occurs in valvular heart disease.

Left Ventricular Stroke Work	43 - 56	$LVSW = \dfrac{(\overline{CI} \times \overline{BP})\ 13.6}{P}$

LVSW
$gm \cdot M \cdot M^{-2}$
Work performed by left ventricle to pump the stroke volume at the aortic pressure.

Total Peripheral Resistance	1300 - 2900	$TPR = \dfrac{(\overline{PB} - \overline{CVP})\ 79.9}{CI}$

TPR
$dyne \cdot sec \cdot cm^{-5} \cdot M^{2}$
Measure of the resistance in the systemic circuit through which blood is pumped.
Elevated in cardiogenic shock, hypertension, hypovalemia. Decreased in septic
shock, vasodilation.

Right Ventricular Stroke Work	6 - 10	$RVSW = \dfrac{(\overline{CI} \times \overline{PA})\ 13.6}{P}$

RVSW $_{M^{-2}}$
$gm \cdot M \cdot$
Work performed by right ventricle to pump its volume at the pulmonary artery
pressure.

Pulmonary Vascular Resistance	100 - 240	$PVR = \dfrac{(\overline{PA} - \overline{PAw})\ 79.9}{CI}$

PVR
$dyne \cdot sec \cdot cm^{-5} \cdot M^{2}$
Measure of the resistance of the pulmonary circuit. Elevated in pulmonary
thromboembolism, congestive heart failure, valvular heart disease.

Parameter & Units	Normal Range	Derivation

Arteriovenous Oxygen Difference   3.1 - 5   A-V Diff= Bound + Dissolved
A-V Diff   arterial oxygen)-(Bound +
Vol%   Dissolved Venous oxygen)

Measure of the difference in oxygen content in arterial and venous blood. Elevated in cardiogenic shock, pulmonary thromboembolism. Decreased in septic shock.

Oxygen Consumption   110 - 155 $QO_2 = \dfrac{(CI)\ (A\text{-}V\ Diff)\ (10)}{Body\ Surface\ Area}$
QO_2
$ml \cdot min^{-1} \cdot M^{-2}$

Oxygen extracted by the tissue from the arterial blood supply.

Veno Arterial Admixture   3 - 9 $Qs/Qt = \dfrac{Calv\text{-}Cart}{Calv\text{-}Cven}$
Qs/Qt   %

That fraction of total blood flow not oxygenated in the lungs. Elevated in cardiogenic shock, septic shock, pulmonary thromboembolism.

Serum Lactate   6 - 18   Quantitative measurement of
Lactate   lactate level in blood.
mgm%

Indicator of hypoperfusion and anaerobic metabolism. Elevated in cardiogenic shock, pulmonary thromboembolism, septic shock.

Arterial Base Excess   0 + 2   Calculation is based upon
BEa   algorithms related to pH,
    -1 PCO_2, hemoglobin, plasma
$meq \cdot L$ proteins (after Thomas, L.J.).

Indicator of the degree of acid-base imbalance.

In vivo P_{50} 26   Defines the partial pressure
mmHg   of oxygen at which 50% of

hemoglobin is saturated under in vivo conditions of pH, pCO_2 and temperature. Calculation is based upon simultaneous measurement of oxygen tension and saturation during blood gas analysis of a venous blood sample.

In vivo P_{50} is altered in disease states (sepsis, cirrhosis), neonates (HbF) and with hemoglobinopathies (Hb Kansas, Hb Chesapeake). Recalculation of oxygen data including A-V diff., oxygen consumption and veno-arterial admixture is performed utilizing the value for in vivo P_{50} and noted on the data format by*_____*.

LVSW-Mixing Volume Plot   Based upon analysis of the indicator dilution curve as a gamma variate function (Thompson,H.J. et al), a mixing volume, V, related to left ventricular ejection fraction may be defined and used to estimate myocardial performance.

REFERENCES

1. JD COHN, P. ENGLER, L DEL GUERCIO: The Automated Physiologic
    Profile. Critical CAre Med. 3,51 (1975)

2. M. KLAIN, S. HIRSCH, A. SLADEN: Automated Physiologic Pro-
    file in the Evaluation of Critically Ill Patients.
    Critical Care Med., 4:124 (1976)

3. JD COHN, ET AL: Physiologic Profiles in Circulatory Support
    and Management of Critically Ill. Journal Am. College
    of Emergency Physicians: 6,479 (1977)

THE CONTINUOUS MONITORING OF TISSUE pH IN CRITICALLY ILL ADULTS

Shyam V. S. Rithalia, M.Sc., Peter Herbert, M.B., B.S.,
Jack Tinker, B.Sc., M.R.C.P., F.R.C.S.

Intensive Therapy Unit
Middlesex Hospital
Mortimer Street, London. W.1

INTRODUCTION

Rapid changes in arterial pH are common in critcally ill patients and may only be observed if frequent blood samples are taken.
This has obvious disadvantages and so recently there has been in-
creaseing interest in the development of continous measuring techniques. Various studies have already been published on the monitoring of skeletal muscle pH (pHm) and its relationship to tissue
perfusion.(1-5) Animal and human studies have shown that pHm varies
directly with arterial blood pH (pHa). That pHm varies directly
with arterial blood pH (pHa), and that pHm decreases in hypovolaem-
ic states and in states of diminished peripheral perfusion of other
causes.(6-8) The pH probes that were used in this work were,
however, cumbersome and difficult to position.

Stamm et al(9,10) have recently designed a miniaturized glass
probe for the measurement of tissure pH (pHt) and it has been developed further by Roche* (Fig 1). The output from the electrode
is connected to a chart recorder. This system is able to monitor
subcutaneous tissue pH continously and the insertion of the electrode into the patient is simple and non-traumatic. Details of the
structure and the performance of the electrode have been given by
Mindt.(11) Investigations into the relationship between pHt and pHa
have been carried out by several workers on animals(12) and on neonates during delivery,(13-15) but to date no information is available for adults.

The purpose of this preliminary study was, therefore, to determine the relationship between pHt and pHa in critically ill
adults.

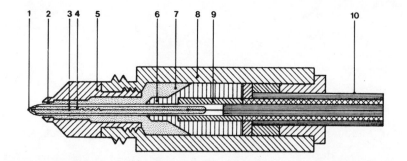

1.   Photograph   and   cross-section   of   tissue   pH
probe.

1 = pH sensitive tip	2 = liquid junction
2 = Ag/AgCl electrode	4 = internal reference fluid
5 = removable tip	6 = Ag/AgCl reference electrode
7 = reference fluid	8 = plastic body
9 = electrical insulation	10 = coaxial cable.

MATERIAL AND METHODS

Twenty-six adults were studied and their ages ranged from 29
to 80 years. Eighteen were studied during recovery from open heart
surgery and ten of these were also studied during cardio-pulmonary
bypass. Prior to use the electrode tip was sterilized in Cidex
(glutaraldehyde) for at least ten hours and then was thoroughly
rinsed with sterile distilled water. The probe was prepared with a
reference solution and then calibrated with fresh buffer solutions
(pH 7.0 and 7.40) according to the manufacturers' instructions.
The shoulder or chest just below the clavicle was used for the monitoring site as these areas were thought to be the most convenient
for the nursing staff and comfortable for the patients. The electrode was placed perpendicular to the skin with its tip penetrating
about four millimetres into the subcutaneous tissue. A skin temperature electrode was placed approximately two centimetres away
from the pH electrode and a rectal probe was inserted to measure

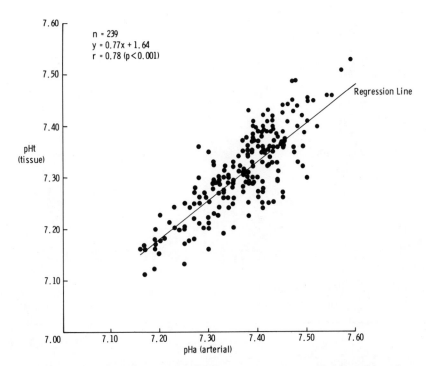

n = 239
y = 0.77x + 1.64
r = 0.78 (p < 0.001)

Regression Line

pHt
(tissue)

pHa (arterial)

2.   Comparison of simultaneous recordings of subcutaneous   tissue   pH and arterial blood pH (regression
line and correlation coefficient).

core temperature. The pHa measurements were made on a Corning 165
blood gas analyser immediately following withdrawal of blood samples from the radial or femoral arteries.

RESULTS

Positioning of the electrode was easily accomplished in less
than one minute and was not uncomfortable for the patients. The
recording stabilized in about three minutes following application
of the electrode and the drift in calibration of the system never
exceeded 0.03 pH units. The electrode functioned satisfactorily in
each case and the tracing was not affected by the movement of the
patient during normal nursing care.

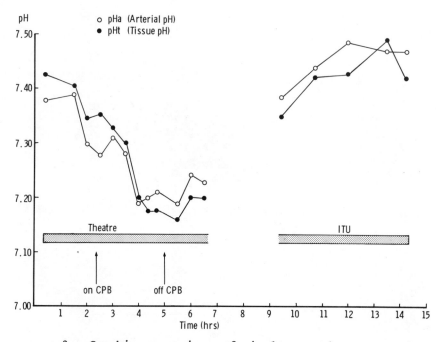

3. Graphic comparison of simultaneously measured
arterial pH and tissue pH.

Two hundred and thirty-nine observations were made on 26 patients and the correlation coefficient for all measurements was 0.78
(Fig 2).   The value, however, varied greatly from patient to patient being higher than 0.9 in six patients, less than 0.7 in five
patients and indicating an extremely poor correlation (r   0.5)   in
another five patients.   In all cases a change in pHa was reflected,
after a time lag, by a similar change in pHt. The measurement of
pHt was generally 0.02 to 0.08 pH units lower than pHa.   Examples
of the relationship between pHt and pHa in individual patients are
shown in the following figures.

Figure 3 shows a comparison between pHt and pHa in a 63   year
old patient who underwent a triple coronary vein graft operation.
The tracing shows a close relationship between pHt and pHa and it
demonstrates a fall in these measurements during cardiopulmonary
bypass (CPB).   Both pHt and pHa rose as the patient recovered . in
the Intensive Therapy Unit (ITU).

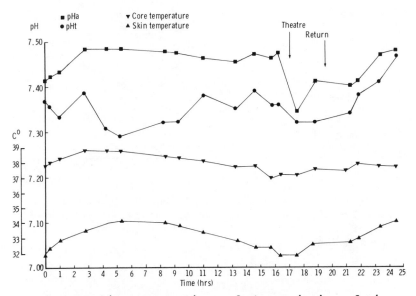

4. Graphic presentations of the variation of tissue pH with arterial pH, core temperature and skin temperature.

Figure 4 shows a 24 hour recording of pHt and pHa taken from a 45 year old patient recovering from a double coronary vein graft operation; skin and core temperatures are also shown. At 1700 hours the patient was taken back to the operating theatre for relief of cardiac tamponade and he returned to the ITU at 1930 hours. Prior to re-operation the two pH measurements were not similar and so the measurement of pHt would have given very poor information regarding the arterial pH of the patient, however, following the relief of the tamponade the pHt and the pHa rose in a similar manner as the patient improved.

Figure 5 shows a continuous recording of pHt with intermittent pHa measurements obtained from a patient who eventually died from lactic acidosis. It shows how pHa was maintained by large infusions of sodium bicarbonate and it illustrates the time lag between a change in pHa and a change in pHt. It explains why a poor correlation between pHa and pHt may be observed at any instant; for example, a close relationship is seen at 40 minutes but a wide discrepancy exists at 80 minutes.

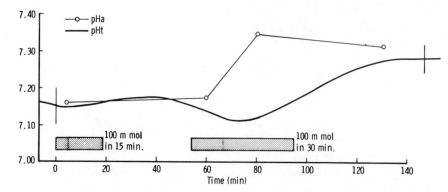

5. Part of a continuous tissue pH recording and intermittent arterial pH measurements. Note the effects of sodium bicarbonate infusion, and the time lag between a change in pHa and pHt.

DISCUSSION

A good correlation between tissure pH and arterial pH has been observed in neonates,(10,15) however, this should not lead us to suppose that they are one and the same. Tissue pH can be a very accurate reflection of arterial pH as seen in Figure 3, but it is also affected by the temperature of the skin and the state of local tissue perfusion. Figure 4 shows how a poor correlation may be obtained in patients with poor tissue perfusion and how the difference between arterial pH and tissue pH provides an index of tissue perfusion which is currently being studied further. The results of this preliminary study are encouraging and suggest that the measurement of tissue pH may be of value in the management of sick adults.

ACKNOWLEDGEMENTS

We wish to thank the surgeons and physicians who permitted us to study their patients, Dr. Mindt for his kind co-operation and Miss Julie Ray for her secretarial services.

*F.Hoffman La Roche Co. Ltd, Basle, Switzerland.
 Supplied by Kontron Medical and Laboratory Systems

REFERENCES

1. SMITH, R.N., ET AL: Surgery, Gynaecology and Obstetrics, 1969,
        128, 533.

2. BERMAN, I.R., LEMIEUX, M.D., AND AABY, G.V.:   Surgery, 1970,
        67,507.

3. COUCH, N.P., ET AL: Annals of Surgery, 1971, 173, 173.

4. GAZZANIGA, A.B., ET AL: Annals of Surgery, 1972, 176, 757.

5. O'DONNELL, T.F., JR,:   Lancet, 1975, 2, 533.

6. LEMIEUX, M.D., ET AL: Surgery, 1969, 65, 457.

7. FILLER, R.M.   AND DAS, J.B., Pediatrics, 1971, 47, 880.

8. FILLER, R.M., DAS, J.B., AND ESPINOSA, H. M.:   Surgery, 1972,
        72, 23.

9. STAMM, O, ET AL: Gynaekologische Rundschau, 1974, 14, Suppl.
        No 1, p. 28.

10. STAMM, O., JANACEK, P., AND CAMPANA, A.:   Acta Gynecologica,
        1975, 26, 265.

11. MINDT, W.:   First International Workshop on Continuous Tissue
        pH Measurement in Obstetrics, Heidelberg, 10 March, 1978.

12. HOCHBERG, H.M., ET AL: Data on file at   Roche   Medical   Elec-
        tronics Inc., Cranbury, New Jersey, 1977.

13. JANACEK, P., ET AL: Geburtshilfe und   Frauenheilkunde,   1975,
        35, 511.

14. STAMM, O., MOLLER, W., AND MAUSER, H.:   Symposium Sur La   Sur-
        veillance Foetale, Paris, 12 - 13 March, 1976.

15. STURBOIS, G., ET AL: American Journal of Obstetrics and Gyne-
        cology, 1977, 128, 901.

ALPHA IONISATION FOR CONTINUOUS COMPUTERIZED CARDIORESPIRATORY MONITORING DURING MECHANICAL VENTILATION

S. Jeretin a, Z. Milavc b,
M. Bratanic c, S. Steen d,
R. Smith d

a Centralna anestezijsko reanimacijska sluzba,
    Klinični center v Ljubljani
b Inštitut Jožef Štefan Ljubljana
c Oddelek za medicinsko elektroniko
    Kliničnega centra v Ljubljani
d Department of Anesthesiology
    Los Angeles County Harbor-UCLA Medical Center
    1000 W. Carson street
    Torrance, CA 80509 U.S.A.

Alpha ionisation pneumotachography (1) has been shown to be suitable for measuring respiratory flows in patients on mechanical ventilators. It is based on the following principles: Americium (Am241), an alpha source, is used to ionise gas molecules passing through a measuring tube-sensor. The concentration of ions above the alpha source depends on the recombination coefficient. These ions are carried away by the gas flow. Downstream from the alpha source there is an electrode. The resulting collector current from the electrode is in linear proportion to the flow (Figure 1).

Humidity in the form of water vapor does not affect the measurement, temperature changes affect the measurement in the range of 0.16% per 1° C. A maximum difference of + 2% exists when oxygen (100%), nitrogen (100%), or carbon dioxide (5%) in oxygen is compared to air (2). After calibration, the sensor remains accurate over long periods of time; in our hospital, over 12 years. For monitoring ventilation on mechanically ventilated patients, the sensor is generally heated to 50° C; in special cases (e.g. exercise tolerance tests), to a higher temperature.

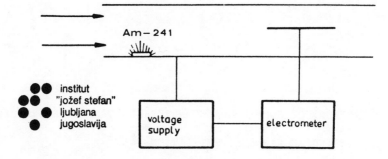

Fig. 1 Block circuit diagram of unidirectional alpha ionisation gas flow sensor.

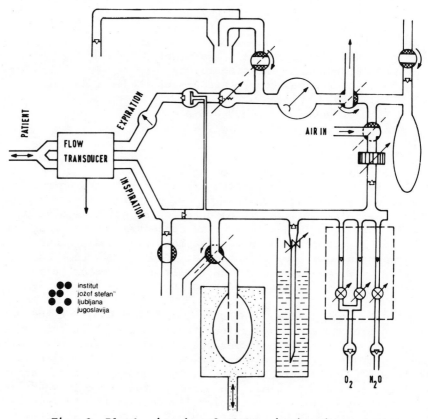

Fig. 2 Block circuit of alpha ionisation sensor in the patient-system (with an Engstrøm respirator).

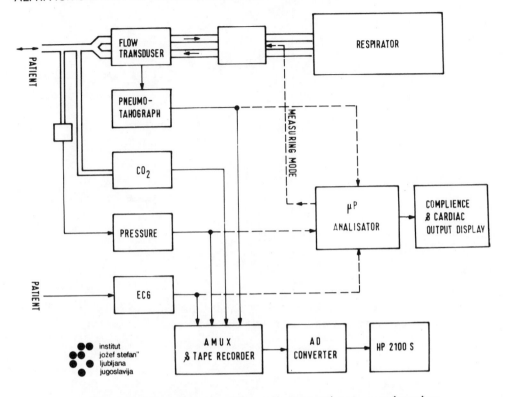

Fig. 3 Flow diagram for computerized monitoring of the cardiorespiratory function during mechanincal ventilation.

The sensor is attached, to the patient system of the ventilator (Figure 2), in a vertical position in order to prevent collection of condensed water on the alpha source. For more extensive monitoring, airway pressure, FECO2, FIO2 are measured simultaneously at the tracheostomy tube adaptor. Analog signals from bedside units are then fed through an analog multiplexer via an AD converter to a Hewlett-Packard 2100 S computer (Figure 3). Respiratory monitoring is based on continuous measurement and evaluation of respiratory flow, rate, tidal volume, minute ventilation airway pressure, fractional concentrations of O2 and CO2 as well as VCO2 and VO2. Pulmonary mechanics are monitored off-line at intervals of several hours. Compliance, airway resistance and the time constant are computed using appropriate equations (Figure 4). The flow pressure and volume pressure relationships as well as the pressures due to resistance and compliance permit the generation of the corresponding curves for resistance (R), compliance (C) and the time constant (T) for a single respiratory cycle.

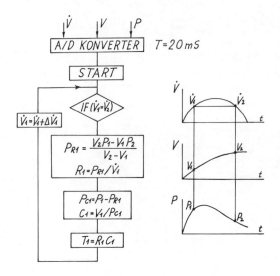

Fig. 4 Flow sheet with equations for airway resistance and compliance (using flow, volume and pressure relationships).

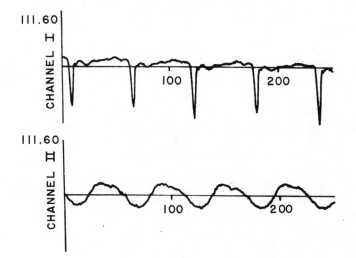

Fig. 5 Pulmonary cardiac capillary aerogram (PCA) as sampled by computer.

Alveolar ventilation and alveolar volume are measured off-line, at intervals of several hours, by means of arterial blood gas analyses.

Simultaneous (continuous) and noninvasive monitoring of the cardiac function, using the alpha ionisation principle, is based on the displacement of air from lung - which is caused by every heartbeat. These displacements of air, pulmonary cardiac (capillary) aerograms (PCA), appear on the pneumotachographic curve at the same frequency as the ECG, and can be amplified, recorded and seperated from the respiratory flow curve by the computer (Figure 5) and a special measuring mode of the ventilator. This method requires the simultaneous recording of ECG. Cardiac function is at present being monitored at intervals in such a manner that a series of four to ten cardiac aerograms are recorded and processed; the mean of these curves being displayed as the curve representing the cardiac aerogram. The computer also estimated the volume of the displaced air per heartbeat, and Figure 6 represents a patient with 'shock lung' whose circulation is in failure and Figure 7 graphically demonstrates the change that has occurred when the same patient has recovered from this circulatory failure.

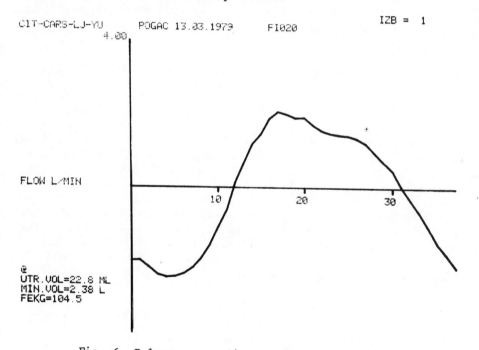

Fig. 6 Pulmonary cardiac capillary aerogram (PCA) of a patient with 'shock lung' in circulatory failure, on an Engstrom respiratør.

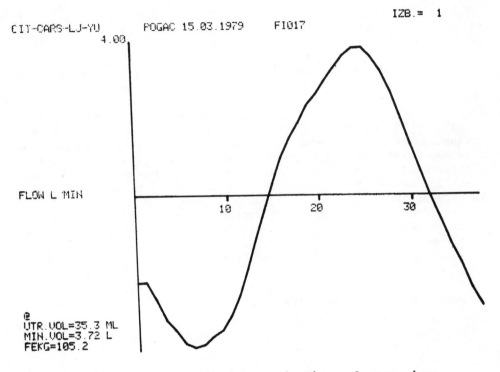

Fig. 7 The same patient as in Figure 9, two   days
later, when his condition is such that he is no longer
in circulatory failure.

The use of alpha ionisation offers some benefits over methods
using measurements of differential pressure. It is accurate, reliable and free of artefacts due to pressure oscillations. In ventilated patients, where fractional concentrations of gases in the patients systems are changed frequently and the gas mixtures are humidified and heated, this method for measuring gas flow is extremely
practical. Using high levels of amplification, we can obtain arte-
fact-free cardiac aerograms.   To measure the pulmonary capillary
flow and cardiac output of ventilated patients, the N20 method (3)
is not readily feasible; whereas this alpha ionization monitoring
technique is readily applicable with only minor modifications of
the respirator being used.

REFERENCES

1. JERETIN, S., HRIBAR, M., AND MILAVC, Z.:   A new control and
       trigger unit for ventilators.   Acta Anaesth Scandinav
       suppl. XXIV: 127-132, 1966.

2. JERETIN, S.,   MARTINEZ,   L.R.,   TANG, I.   P.,   AND ITO, Y.:
       Pneumotachography by the ionization principle - a new ap-
       proach.   Anesthesiology 35(2):218-223, 1971.

3. WASSERMAN, K. AND COMROE,JR, J.H.: A method for estimating
       instantaneous pulmonary capillary blood flow in man. J.
       Clin Invest 41:401-410,1962.

CLINICAL BASIS AND USE OF AN AUTOMATED ICU TESTING SYSTEM +

Richard M. Peters, M.D., John E. Brimm, M.D.
Clifford M. Janson

Department of Surgery/Cardiothoracic
University of California Medical Center
San Diego, California

+ Supported by USPHS Grants GM 17284, GM 25955 and HL 13172, and ONR Contract N00014-76-C-0282.

Tools for monitoring respiration to prevent accidents and to detect catastrophic failures are still in their infancy. These tools are essential, however, because the relative danger of accidental airway disconnection or ventilator malfunction increases directly with a patient's dependence on ventilator support. The most dependent patients have drug or disease induced muscle paralysis and cannot develop spontaneous ventilatory force. At the other extreme, patients on intermittent mandatory ventilation (IMV) require only a small number of breaths per minute to supplement their own ventilatory effort. A continuous monitor of expiratory volume, respiratory rate, and airway pressure is needed to prevent catastrophic failures. Unfortunately, this full mix has rarely been available, principally because of the unreliability of airflow sensors.

Most ventilators in use today have only crude monitors, using aneroid manometers and relatively inaccurate volume measuring devices. Some recently available respirators have adequate airflow, volume, and airway pressure sensors to monitor respirator breaths. The advance in respirator therapy made by IMV makes these monitors less useful since they do not analyze a patient's spontaneous breaths. A monitor that can separate IMV and spontaneous tidal volume and respiratory rate will be extremely useful. Hopefully, new flow sensors, such as that designed by John Osborn, (1) and the advances in microcomputer technology will stimulate the development of ventilator monitors suitable for IMV. However, these monitors will not provide a full set of information for assessing respiratory function.

To provide a more complete information set, we have focused a major effort on developing a sophisticated set of programs and the necessary hardware to allow complex cardiopulmonary testing for the ICU patient. (2,3) Since pulmonary function has a long time constant of change, frequent sophisticated testing is unnecessary. Once or twice a day is adequate in most patients. For decision making, pulmonary function tests must be coordinated with analyses of blood gases and cardiac function. In this paper, we will describe how we use our testing system to improve decision making.

IMPLICATIONS OF IMV

Our system was originally developed for analysis of the problems associated with initiating, adjusting, and discontinuing controlled ventilation. We now use IMV almost exclusively. Controlled ventilation can be considered a special case of IMV, where the IMV rate and tidal volume have been increased to the point that a patient's spontaneous effort is suppressed. However, spontaneous effort is not prevented as it is with controlled ventilation. To cover the wide range of IMV and not just the special case of controlled ventilation requires different testing systems and analytic logic.

When analyzing breathing for a patient on controlled ventilation, testing systems evaluate the passive ventilation of the lungs by a machine. If airway pressure is used for measuring compliance, resistance, or work, then accurate analysis depends on the complete suppression of patient effort. Patient effort opposing the ventilator will increase work and lower compliance, while patient effort synchronous with the ventilator will decrease work. These effects are exaggerated when IMV is used, and we have found that measuring compliance using airway pressure is worthless. Lung mechanics can be evaluated using an esophageal balloon to measure transpulmonary pressure.

The use of IMV recognizes that the bellows function of the chest cage is as important as the mechanical and gas exchange functions of the lungs. The relationship between a patient's ability to support his own ventilation and the ventilator's supplementation provides important information for functional analysis. A patient on IMV is at lower risk if support is temporarily interrupted to test ventilatory effort. Because IMV also allows stepwise withdrawal of respiratory support, analysis of the response to stepwise decreases provides added useful information. This paper will concentrate on the use of information from the testing system for patients on IMV.

PHYSIOLOGICAL BASIS

To make intelligent decisions, one must identify the ultimate goals.(4) The purpose of the lungs is to excrete CO2 and acquire oxygen, both of which are transported by blood that is pumped by the heart. A measure of the success of this system is the amount of oxygen available to the tissues -- the product of cardiac output (Qt) and arterial oxygen content (CaO2). The first step in decision making must be to appraise the relative importance of Qt and CaO2 in optimizing available O2 and in turn cardiopulmonary function. The concept that optimizing Qt should be the goal for adjustment of cardiac function is easy to understand. To focus on CaO2 as the goal for pulmonary optimization is less obvious and, in fact, CaO2 cannot be used alone.

Deciding on priorities for treatment requires understanding of the basis of pulmonary dysfunction. West and his disciples (5) have shown with brilliant studies that most types of clinical disorders of gas exchange have as their basis ventilation-perfusion incoordination. The fact that the major dysfunction of the lungs is incoordination of ventilation and perfusion permits one to define the interaction between cardiac and pulmonary function and define priorities for optimization. The basis for the definition of priority can be found in the shunt equation. The analysis of this equation not only leads to priority between CaO2 and Qt but also priority for treatment to enhance both.

The shunt equation defines the factors that alter CaO2.

$$Q_s/Q_t = \frac{C_AO_2 - CaO_2}{C_AO_2 - CvO_2}$$

where Q_s/Q_t = fraction of blood going to unperfused alveoli. This can be used to define anatomically unperfused alveoli or a physiologic equivalent to unperfused alveoli.

CAO2 = oxygen content of blood draining well ventilated alveoli

CaO2 = oxygen content of arterial blood.

CvO2 = oxygen content of mixed venous blood.

Transposed to define CaO2, the equation is CaO2 = CAO2 - Qs/Qt (CAO2 - CvO2). CaO2 will be decreased by a fall in CAO2, a rise in Qs/Qt and a fall in CvO2. Alterations in CAO2 and Qs/Qt will result from changes in pulmonary function. CvO2 is most affected by a change in cardiac output; a fall in cardiac output lowers CvO2. Therefore, a fall in cardiac output will also lower CaO2. Quantitative analysis of the relative importance of a fall in Qt versus a fall in CaO2 on available oxygen shows dominance of cardiac output. Optimal care must therefore focus first on maintaining cardiac output.

Analysis of the shunt equation also directs the primary goal for control of pulmonary function. If blood is perfusing unventilated or hypoventilated alveoli, then hyperventilation of the remaining alveoli might compensate. Unfortunately, this compensation, as far as CAO2 is concerned, is limited because of the shape of the O2 dissociation curve. Above a PaO2 of 95, little oxygen is added to blood and the rate of increase in CaO2 relative to PaO2 falls off rapidly after a PaO2 of 65 is reached. Thus, hyperventilation of aerated alveoli can have only limited effect in correcting for unventilated alveoli, and PaO2 is a poor index of the adequacy of alveolar ventilation. However, general alveolar hypoventilation is a primary cause of fall in PaO2 and must be prevented. Consideration of the adequacy of alveolar ventilation illustrates how keying on available oxygen forces consideration of the lungs' excretory function, excretion of carbon dioxide.

In contrast to PaO2, PaCO2 is a direct index of the level of alveolar ventilation. The CO2 dissociation curve over the physiologic range is essentially a straight line, so hyperventilation of ventilated alveoli can correct for lack of CO2 elimination in others. By keying on the level of PaCO2, one can assess the adequacy of alveolar ventilation and the elimination of CO2 and be certain that one cause of CaO2 depression is eliminated. The earliest step in cardiopulmonary optimization is to assure adequate alveolar ventilation to maintain PaCO2 at a level that produces a pH as near 7.40 as possible. The goal, optimize available oxygen, thus assures optimal oxygen acquisition, CO2 elimination and transport, as well as pH control.

CLINICAL ALGORITHMS

Analysis is initiated when blood gas measurements become available. PaO2 is not used for primary adjustment of alveolar ventilation since its abnormalities are multifactorial, only one of which is inadequate alveolar ventilation. The first step is to evaluate the acid-base disturbance (Fig. 1). An acid-base program(2) provides immediate identification of the type of acid-base disturbance. If respiratory acidosis is present appropriate adjustment of alveolar ventilation must be achieved prior to assessment of the adequacy of oxygen delivery.

PRIMARY ACID BASE ANALYSIS

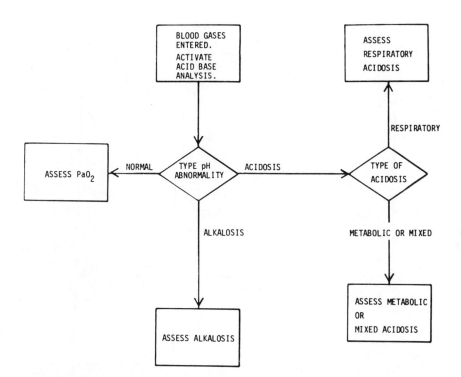

Figure 1.   Initial logic for determining treatment priorities based on entry of blood gases.

If the disturbance is diagnosed by the acid-base program as respiratory acidosis (Fig. 2) and the patient is not on a ventilator, the physician must decide whether ventilator support is indicated.   A $PaCO_2$ of 50 is accepted because with no metabolic component, circulation is presumably good and elevations of $PaCO_2$ above 40 are common in post-operative patients. If the patient is receiving ventilator assistance, the alveolar ventilation should be adjusted to achieve a normal $PaCO_2$ of 40 by increasing rate or tidal volume (TV). We prefer in the IMV mode of ventilation to get TV to at least 15 cc/Kg before increasing the rate. Restrictions on increasing TV may be necessary if peak and/or mean airway pressure are elevated.

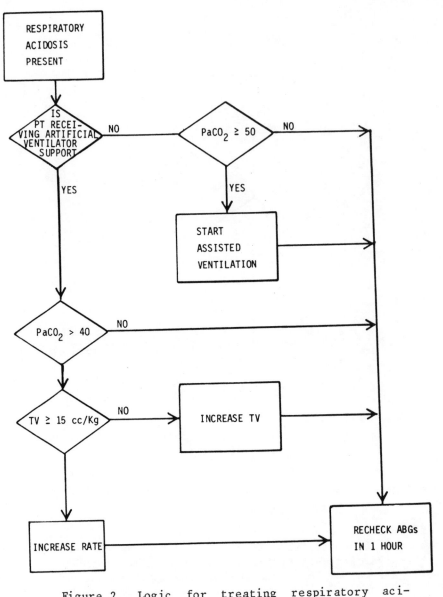

Figure 2. Logic for treating respiratory acidosis.

The alteration of alveolar ventilation is the primary response
to preserve normal pH. For mixed or metabolic acidosis (Fig. 3),
the level of PaCO2 must be assessed to ascertain whether there is
significant respiratory compensation for the metabolic acidosis --
if there is no respiratory compensation, the advice is to give ventilator support if assisted ventilation is not being provided, or

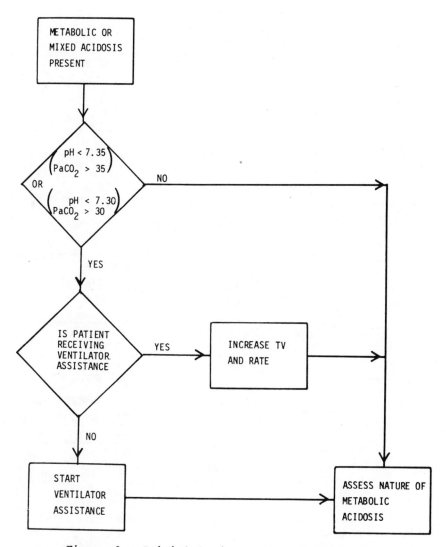

Figure 3. Initial logic for treating mixed or metabolic acidosis.

to increase alveolar ventilation if the patient is on a ventilator. This assures that the primary physiologic system for correcting pH is activated if metabolic acidosis is present. Once respiratory compensation for low pH is assured, one can proceed to determine the cause of metabolic component of the acidosis (Fig. 4).

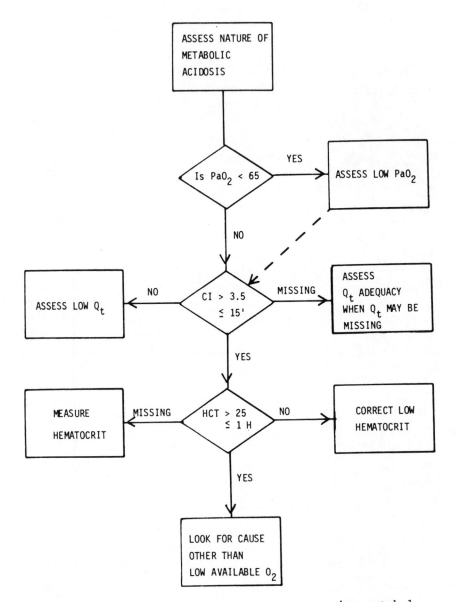

Figure 4.   Subsequent logic for assessing metabolic acidosis.

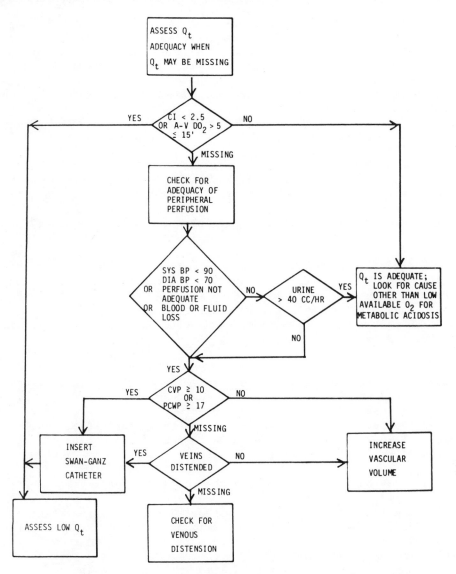

Figure 5. Assessment of the adequacy of cardiac
output when direct measurement may not be available

    Failure to protect cardiac output is one of the most frequent
and devastating errors made when the physician is focusing on pulmonary dysfunction.(6) We concentrate on the most common cause of
metabolic acidosis during the postoperative or post-injury period,
decrease in available oxygen. The first step is to ask for information from the blood gas entry which initiated the systems analysis. PaO2 depression below 65 is a possible direct cause of the

low available O_2 and must be assessed as top priority. The logic tree appears to disregard Qt if PaO_2 is depressed. This is not the case since, as indicated in the dotted line connecting the PaO_2 depression box to the Qt decision, the first step in the analysis of depressed PaO_2 is to assess the adequacy of Qt. If a direct measurement of Qt is present and cardiac index (CI) is above 3.5, the hematocrit must be checked. If it is above 25 and PaO_2 above 40, then the metabolic acidosis is not due to low available O_2 and other causes of metabolic acidosis must be considered.

The treatment of low Qt has priority above the treatment of low CaO_2 because analysis (7) shows that Qt has greater quantitative effect on available oxygen by the direct effect apparent from its calculation as a product of CaO_2 and Qt, and because a fall in Qt lowers CvO_2. As the shunt equation shows, a fall in CvO_2 results in lower PaO_2 at the same shunt level.

Therefore, in the description we will deal with this problem first, though in fact the clinician would attempt to deal simultaneously with depressed Qt and CaO_2 if they were both present. Since many patients will not have directly determined cardiac output, one must make an assessment of adequacy of Qt when direct determination is not available.

Figure 5 depicts the low cardiac output decision tree. In priority of usefulness, we use arterio-venous oxygen difference (A-V DO_2), systemic blood pressure, observer evaluation of peripheral perfusion, wedge pressure, and central venous pressure. The latter two prove useful primarily in analyzing the cause, but, if low, volume loading should be used to see if the other indicators of low Qt move in a positive direction.

The A-V DO_2 is included with the cardiac output because it will rarely be available when Qt is unavailable. The missing data logic is used to make a decision without direct measurement of Qt. We check for the presence of metabolic acidosis, evidence of fluid or blood loss and the level of blood pressure. If none of these are found to indicate low Qt, we check for urine output and, if low, proceed to evaluate venous filling. If urine output is adequate, the advice is given to look for another cause for metabolic acidosis. If there is evidence of vein distension or directly measured high filling pressures, cardiac tamponade is likely and a Swan Ganz catheter should be inserted to measure Qt and wedge pressure. If there is no vein distension, a fluid challenge should be tried, followed by a recheck of perfusion and blood pressure, etc.

When pulmonary capillary wedge pressure or left atrial pressure is greater than 14, blood infusion does not necessarily increase Qt. (8) If volume deficit is not the cause of low Qt, the logic must then differentiate between primary excessive afterload and/or impaired myocardial contraction.

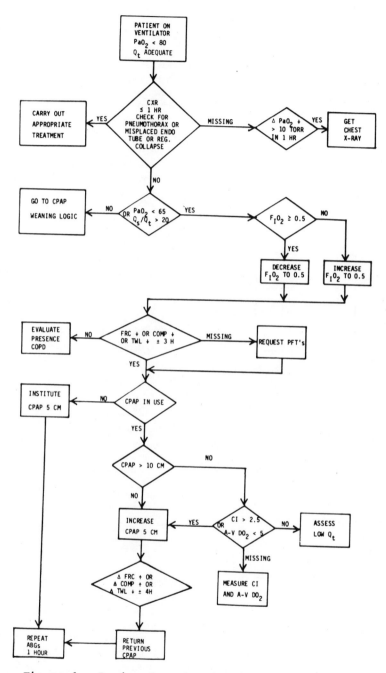

Figure 6. Logic for instituting or increasing CPAP.

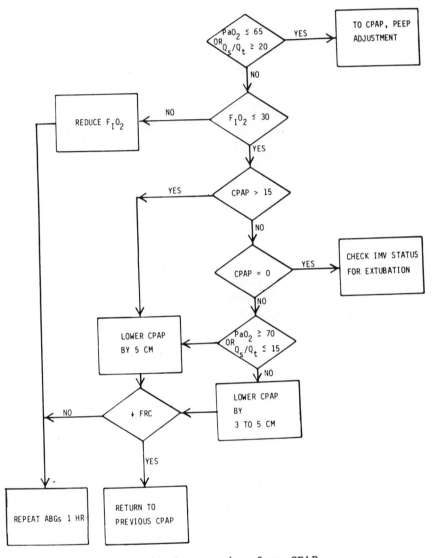

Figure 7.   Logic for weaning from CPAP.

Once adequacy of Qt and alveolar ventilation is assured, we proceed to evaluate the need for institution of CPAP (Fig.6). One must first review the chest x-ray to be sure that a low PaO2 is not caused by pneumothorax, misplaced endotracheal tube or other physical derangements that will respond rapidly to appropriate therapy. If PaO2 65 or Qs/Qt 20, then FIO2 should first be evaluated and adjusted up to 0.5 if above that level to avoid the risk of oxygen toxicity. If Qt is adequate, a low PaO2 can be tolerated and risking further damage to the lungs by oxygen is not warranted.

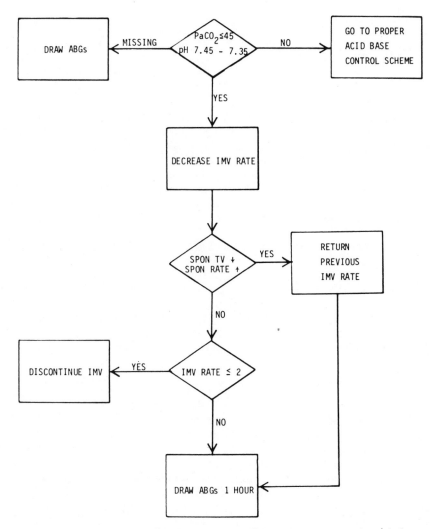

Figure 8. Logic for decreasing IMV rate and tidal volume.

The next step in controlling treatment is to evaluate pulmonary function. If functional residual capacity (FRC) and compliance are not depressed, the patient most likely has obstructive airways disease and bronchodilators should be tried. The more common problem in surgical patients is a low FRC and compliance and an increased TWL.

If FRC or compliance are low or work elevated and CPAP is in use, if CPAP is 10 cm, we would not increase CPAP until Qt is measured and found to be adequate. If CPAP is § 10 cm, we would increase CPAP whether or not direct measurement of CI is available. We check to see if the increase in CPAP raised FRC and raised compliance and/or lowered work. If not, we return to the previous settings. If yes, we repeat blood gases in one hour, which will start our cycle of analysis again.

If PaO2 is above 65 and Qs/Qt is below 20, we begin the decisions to lower CPAP (Fig. 7). First we lower FIO2 to reach 0.3 before adjusting the CPAP level. If CPAP is above 15, we proceed directly to try to lower CPAP. If CPAP 15, PaO2 70 and Qs/Qt -15, we lower CPAP. The FRC is checked and if it falls when CPAP is lowered, the PaO2 will fall later so we move CPAP back. If FRC does not fall or the measurement is not available, in one hour blood gases are drawn and the analysis begins again.

The control of IMV is not done with PaO2 but by the PaCO2 and pH and spontaneous rate and tidal volume (Fig. 8). We use the previously described alveolar ventilation controls. The logic tree for lowering IMV merely adds on the left-hand part of the diagram the weaning routine which lowers the assistance provided by IMV. If spontaneous rate and TV rise as a result of lowering IMV, we return to the previous level as we have found that this warns of a later rise in PaCO2.

CONCLUSION

In the control of IMV and CPAP, where we have the most investigative experience, the numerical values for keying responses are not precisely defined. We are in the process of continued evaluation for various patient populations. A number of clinical situations in trauma and cardiac surgical patients will restrict the depicted decision logic. For example, the head injury patient requires hyperventilation to induce cerebral vasoconstriction, and CPAP may raise venous and intracranial pressure, particuarly if the lungs are compliant. In crushed chest patients, we now avoid prolonged ventilator support and concentrate on pain control, so that measures of spontaneous ventilation are critical. (9)

The most important fact we have learned is that the automated information system provides no useful aid if used improperly. In too many instances, the complexity of the information overwhelms the ability of the clinical staff to use it for intelligent deci- sion making. To deal with the consequences of drowning the clini- cal staff in raw data, we are now focusing our efforts on program- ming the decision logic to provide appropriate advice, not massive tables of data. The software tools required for undertaking this task have been described in a separate paper.(10)

ABBREVIATIONS

A-V DO2	Arterio-venous oxygen difference
CaO2	Arterial oxygen content
CAO2	Oxygen content of blood draining well ventilated alveoli
CvO2	Oxygen content of mixed venous blood
CI	Cardiac index
CPAP	Continuous positive airway pressure
CVP	Central venous pressure
FIO2	Fraction of inspired oxygen
FRC	Functional residual capacity
IMV	Intermittent mandatory ventilation
PaCO2	Arterial carbon dioxide partial pressure
PaO2	Arterial oxygen partial pressure
PEEP	Positive end-expiratory pressure
Qs/Qt	Intrapulmonary shunt fraction
Qt	Cardiac Output
TV	Tidal volume
TWL	Total work of breathing/liter

REFERENCES

1. OSBORN, J.J.: A flow meter for respiratory monitoring. Crit.
    Care Med. 6: 349-351, 1978.

2. BRIMM, J.E., JANSON, C.M., KNIGHT, M.A. AND PETERS, R.M.:
    Extracting information from blood gas analyses. Proc.
    First Annual International Symposium Computers in Criti-
    cal Care and Pulmonary Medicine, Norwalk, Connecticut,
    May 24-26, 1979.

3. JANSON, C.M., BRIMM, J.E., AND PETERS, R.M.: Instrumentation
    and automation of an ICU respiratory testing system.
    Proc. First Annual International Symposium Computers in
    Critical Care and Pulmonary Medicine, Norwalk, Connecti-
    cut, May 24-26, 1979.

4. PETERS, R.M.: Life saving measures in acute ARDS. Am. J.
    Surgery, in press.

5. WEST, J.B., ANTHONISEN, N.R., GLAISTER, D.H., HUGHES, J.M.B.,
    AND MILIC-EMILI: In J. B. West (Ed.), Regional Differ-
    ences in the Lung. New York: Academic Press, 1977.

6. POWERS, S.R.,JR., MANNAL, R., NECLERIO, M., ET AL: Physiologic
    consequences of positive end-expiratory pressure (PEEP)
    ventilation. Ann. Surg. 178: 265, 1973.

7. HOBELMANN, C.F., JR., SMITH, D.E., VIRGILIO, R.W. AND PETERS,
    R. M.: Mechanics of ventilation with positive
    end-expiratory pressure. Ann. Thorac. Surg. 24:
    68-76, 1977.

8. KOUCHOUKOS, N.T. AND SHEPPARD, L.C.: Postoperative evaluation
    of cardiac function. In A.P. Fishman (Ed.), Heart Fai-
    lure. Washington, DC: Hemisphere, 1978.

9. SHACKFORD, S.R., SMITH, D.E., ZARINS, C.K. ET AL.: The man-
    agement of flail chest. Am. J. Surgery 132: 759,
    1976.

10. BRIMM, J.E., JANSON, C.M., PETERS, R.M. AND STERN, M.M.:
    Computerized ICU data management. Proc. First Annual
    International Symposium Computers in Critical Care and
    Pulmonary Medicine, Norwalk, Connecticut, May 24-26,
    1979.

ORGAN SYSTEM DATA DISPLAY IN AN ICU DATA BASE MANAGEMENT SYSTEM

R.S. Swinney, M.D., J.R. Stafford, M.D.,
P.W. Wagers, M.D., M.W. Davenport, M.D.,
W.S. Wheeler, M.D.

I would like to present to you today our solution to a serious problem that was alluded to by a number of yesterday's speakers, and one that is repeatedly faced by critical care physicians. That is, how to deal most productively with the large volume of information that necessarily accrues to patient in intensive care units. Traditionally, medical data - especially laboratory data - is evaluated by reviewing a source document - for example, a laboratory slip reporting serum chemistries, an SMA 12, arterial blood gases, or a fluid balance sheet showing intake and output details. Some physicians summarize multiple source documents onto a large flow sheet containing all sorts of information. This type of approach can be useful, but what is potentially more valuable at the bedside is to examine the results of those tests, and just those tests, that are applicable to a particular medical problem or disease.

Consider, for example, the problem of non-cardiogenic pulmonary edema (Fig. 1). A reasonable basic set of data needed to assess the status of this problem might include, in addition to the physical examination and chest x-ray, daily weights, intake and output totals, ventilator settings and measurements of respiratory mechanics, arterial blood gases, and perhaps plasma colloid oncotic pressures. Understand that this problem is only one of the hundreds of different diseases or problems that occur in ICU patients; obviously, it would be a monumental task to define the pertinent data set applicable to each individual disorder. Even if such a task could be accomplished and implemented on a computer based data management system, the result would be a massive unwieldy system, incapable of rapid efficient data display.

PROBLEM

NON-CARDIOGENIC PULMONARY EDEMA

PERTINENT DATA

| WEIGHTS | INTAKE AND OUTPUT |
| VENTILATION SETTINGS/MEASUREMENTS |
| BLOOD GASES | ONCOTIC PRESSURES |

Figure 1.

Date Each Entry	Date of Onset	ACTIVE PROBLEMS	Date Resolved
Prob. = 1	4/15/79	ACUTE MYOCARDIAL INFARCTION, S/P CPA, c̄ CHF	*
Prob. = 2	4/15/79	FLAIL CHEST, RIGHT PNEUMOTHORAX	*
Prob. = 3	4/15/79	RESPIRATORY FAILURE	*
Prob. = 4	4/17/79	PNEUMONIA, DIPLOCOCCAL	*
Prob. = 5	1974	CEREBROVASC. DISEASE, S/P CVA	
Prob. = 6	?1970	A.O.D.M	
Prob. = 7	?	COPD	*
Prob. = 8		ETOH ABUSE	

Figure 2.

Now, (Fig. 2) in this partial problem list of a patient recently in our ICU, note that 5 of the "problems" identified by the patients's physician - and here by the asterisks - bear some relationship to pulmonary dysfunction. The data sets needed for evaluation of each of these problems will be similar to some degree. At least there will be a number of pieces of information common to all these problems.

So, rather than examine similar data repeatedly under each problem, as would be done in the usual "problem oriented" approach, a more efficient method of data retrieval and display is to develop rational medical "groupings" of information. We feel that the most useful grouping or categorization of medical data in the ICU is by organ systems.

```
MICU 3                                   16 MAY 1435

    1   SELECT OR ADMIT PATIENT
    2   MOVE THIS PATIENT
    3   CHANGE MONITORING STATUS
    4   DISCHARGE THIS PATIENT
    5   PATIENT IDENTIFICATION
    6   TRAINING AND REFERENCE LIBRARY
    7   LAB AND ICU DATA ENTRY
    8   LAB AND ICU DATA DISPLAY
    9   HEMODYNAMIC DATA ENTRY
   10   HEMODYNAMIC DATA DISPLAY    [A]
   11   HEMODYNAMIC SUMMARY         [B]
   12   ORGAN SYSTEM DATA DISPLAY   [C]
   13   TREND PLOTS
   14   PROGRESS NOTES
   15   STATUS DISPLAY SET UP
   16   PRINTED REPORTS
   17   CURRENT VITALS
   18   MULTI PATIENT VITALS DISPLAY
   19   ARRHYTHMIA STATUS DISPLAY

CHOOSE, OR PUSH GO FOR MORE:
```

Figure 3.

We have implemented this organ system data display concept on a Hewlett Packard 5600A Patient Data Management System — as you can see as item 12 in the list (Fig. 3) of system functions callable by the user. Note that other methods of data display are available — for example, 8, which displays data by source document, and 10, which displays hemodynamics as was illustrated yesterday.

```
MICU 3                                   16 MAY 1447

                PATIENT'S PROBLEM LIST

    1   DATA BASE REVIEW                  ACT
    2   HEMODYNAMIC SUMMARY               ACT
    3   CARDIOVASCULAR TRENDS             ACT
    4   CV TRENDS W/ VASOACTIVE DRUGS     ACT
    5   PULMONARY                         ACT
    6   RENAL                             ACT
    7   LIVER                             ACT
    8   ENDOCRINE                         ACT
    9   HEMATOLOGY                        ACT
   10   NEUROLOGY                         ACT

CHOOSE PROBLEM # OR 0 TO EXIT
```

Figure 4.

We currently have developed data presentation structures for the cardiovascular system – (Fig. 4) items 2, 3, and 4 – and the pulmonary system – item 5 – as well as the renal, hepatic, endocrine, hematologic, and neurologic organ systems. Patient data may be viewed in one of two formats – tabular or graphic.

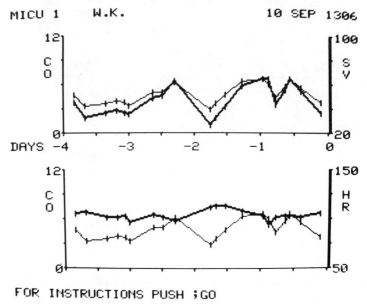

MICU 1   W.K.   10 SEP 1306

Figure 5.

MICU 8   15 MAY 0838

HEMODYNAMIC SUMMARY (1)

	13 MAY 2200	14 MAY 0200	14 MAY 1830	15 MAY 0200
HR	102.	95.	96.	104.
SA-S	132.	102.	102.	72.
SA-D	55.	48.	58.	44.
SA-M	81.	66.	73.	53.
T	38.3	38.3		
PA-S	46.	46.	40.	34.
PA-D	18.	10.	12.	15.
PA-M	27.	22.	21.	21.
PA-W	18.	16.	13.	11.
RA-M	12.	14.	9.	10.
CO	9.3	6.4	6.1	4.8
SV	91.6	67.8	63.5	46.2
LVSW	100.5	60.8	62.8	33.5
SVR	588.	645.	834.	721.
PVR	79.9	74.5	109.2	172.0

FOR INSTRUCTIONS PUSH ;GO

Figure 6.

MICU 3   16 MAY 1437

CARDIOVASCULAR TRENDS

```
->      INVASIVE VITAL SIGNS: ARTERIAL
 2.     INVASIVE VITAL SIGNS: SWAN-GANZ
 3.     NON-INVASIVE VITAL SIGNS
 4.     DETERMINANTS OF CO: SV/HR
 5.     DETERMINANTS OF SV: PA-W/LVSW
 6.     DETERMINANTS OF SV: AFTERLOAD
 7.     CO VS. PRELOAD AND AFTERLOAD
 8.     STROKE WORK: LEFT HEART
 9.     STROKE WORK: RIGHT HEART
10.     CO VS. R & L FILLING PRESSURES
11.     PA-W VS. RA-M AND PA-D
12.     HYPOTENSION: "MECHANISMS"
13.     CO VS AVDO2/PvO2

#  GO SELECTS FRAME   ; GO RETURN TO FRAME
   GO ADVANCE FRAME   - GO NEWER PAGE
0  GO EXITS           . GO OLDER PAGE
```

Figure 7.

This first of two hemodynamic summary frames (Fig. 5) shows measurements and calculated variables at 4 different times, as specified at the top of each column. The measured data includes heart rate, systemic arterial systolic, diastolic and mean blood pressures, temperature, pulmonary artery pressures, right atrial mean pressure, and cardiac output. Derived variables are stroke volume, left ventricular stroke work, systemic vascular resistance, and pulmonary vascular resistance.

We also display plots of cardiovascular variables, such as this one (Fig. 6) showing cardiac output (the thin line) versus heart rate (the thicker line) and cardiac output versus stroke volume for the last 4 days. A large number of graphic displays (Fig. 7) are available for use in evaluation of a patient with an unstable cardiovascular system. The primary concept here is that these displays are applicable to the evaluation of a patient°s cardiovascular status, regardless of which disease, syndrome, or problem is responsible for the abnormal cardiovascular physiology.

Because so many patients in the intensive care unit exhibit abnormal lung function, the pulmonary system data display frames are particularly valuable. This is the list (Fig. 8) of the frames we have generated so far.

```
MICU 8                                  15 MAY 0845

          PULMONARY

   1.     RESPIRATORY FAILURE SUMMARY

   ->     VENTILATION INDICES        (1 OF 4)

   3.     ARTERIAL BLOOD GASES

   4.     MIXED VENOUS BLOOD GASES

   5.     OXYGEN DELIVERY

   6.     OXYGEN-HEMOGLOBIN RELATIONSHIPS

   7.     WEANING CRITERIA

   8.     ACID-BASE STATUS

 # GO SELECTS FRAME    ; GO RETURN TO FRAME
   GO ADVANCE FRAME    - GO NEWER PAGE
 0 GO EXITS            . GO OLDER PAGE
```

Figure 8.

```
MICU 8                                  15 MAY 0845

RESPIRATORY FAILURE SUMMARY

              14 MAY   14 MAY   15 MAY   15 MAY
              1830     2130     0015     0200
    VENTLR     1.                1.
    IMV
    RR-SET    20.               20.
    RR-OBS    28.               24.
    V-T      1000.             1000.
    COMP-S    29.0
    PIP       45.               40.
    PEEP
    FIO2       .99      .99               .99
    PO2       61.      51.               53.
    HGB       14.0     13.2              13.8
    SaO2-C    81.      64.               66.
    CaO2      15.7     11.9              12.7
    PCO2      58.      74.               66.
    PH        7.24     7.15              7.13
    HCO3      24.      25.               21.
```

FOR INSTRUCTIONS PUSH ;GO

Figure 9.

The respiratory failure summary frame (Fig. 9) contains information related to the mechanics of ventilation - respiratory rate,
tidal volume, compliance and airway pressures, as well as data related to lung function as reflected by arterial blood gas values.
By reviewing medical data in this fashion, we are able to assess
the functional derangement of each organ system without being constrained by specific etiologic considerations.

```
MICU 8                                    15 MAY 0845

ACID-BASE STATUS

              14 MAY   14 MAY   14 MAY   15 MAY
              1800     1830     2130     0200
        PH              7.24     7.15     7.13
        PCO2            58.      74.      66.
        HCO3            24,      25.      21.
        BE              -2.      -2.      -6.
        BUN     57.                       54.
        CR-S    2.0                       2.5
        NA-S    145.                      148.
        K-S     4.9                       5.9
        CL-S    109.                      116.
        HCO3-   24.                       27.
        CHO-S   67.                       66.
        CHO-U            0.
        KETO-U           0.
        KETO-S
        LACTAT          26.               52.
        PH-U    6.50

FOR INSTRUCTIONS PUSH ;GO
```

Figure 10.

Finally, I would like to point out that the computer is a necessary element to the organ system data assessment approach because of the large amount of data manipulation involved. Observe
that this display (Fig. 10) of acid base balance includes information which must be brought together from many different source documents - pH, pCO2, calculated bicarbonate and base excess from arterial blood gases; BUN, electrolytes and blood sugar from the
serum chemistry panel; urine sugar and ketones from the S _A record; and lactate from a separate serum panel.

So, in conclusion, I would state that we have utilized the
computer's capabilities of organization and display of data in a
medically logical format, so that data evaluation can be accomplished systematically, thoroughly, and efficiently, thereby contributing to the ultimate goal of more effective patient care.

THE AMORPHOUS DATA BASE IMPLEMENTATION ON MICROPROCESSORS AND RELATION TO NON ALGORITHMIC DIAGNOSIS

Stuart B. Pett, Jr., M.D.; Eddie I. Hoover, M.D.;
William A Gay, Jr. M.D.; Valvanuar Subramanian, M.D.

Department of surgery, Thoracic Surgical Division
New York Hospital/Cornell Medical Center, New York

An Amorphous Data Base (ADB) can be defined as a method of file storage in which the quantum of data, and all identifying labels, are associated as a discrete unit. Such files can be sorted or searched via any of these identifying parameters without disrupting the integrity of the basic information. This type of filing should be contrasted with the Defined Data Base, (DDB), where specific arrays of data are placed into predefined, non-variable, storage areas. Each approach has its own individual strenghts and drawbacks, depending on the requirements presented. If the data is specific, consistent, and complete, such as output from an ECG, an SMA-12, or an Automated Review of Systems, (Ref 6,15), a DDB undoubtedly is of greatest utility in storage and processing. Each variation, though, exception or deletion detracts from those overall advantages.

Clinical medicine, however, does not always lend itself to these rigid constraints. In the long run, the only thing that appears consistent in medicine is its variation. Straight forward, clear cut cases are often an illusion created by 'simplifying generalizations'. An ADB then, would seem to be necessary for a patient management system that attempts to address the problem of comprehensive patient care. It is a method which allows the physician, during the on-going data acquisition process, to determine the relevance of the observations, rather than having rigid priorities assigned at the outset.

A major consideration in choosing between data bases pertains to hardware. Before the microprocessor era, automated data management was restricted to those institutions with considerable resources, regional medical communities, university hospitals, or governmental plans. A number of systems have been developed that employ variations of the DDB. COSTAR, EXETER, Regenstrief Institute, Arizona Health Plan, ARAMIS, CAPO, Oxford Community Health Project, PROMIS, and others, are currently operational. (Ref. 1,3,12,4,8,13,14,16,20). As a consequence, the processing overhead often had to be distributed over many patients. This led to the practice of accumulating specialized data on specific patient subsets. This may occasionally impose some significant restrictions when a patient has multiple diagnoses, tests, or procedures. Given the reality of certain compelling financial implications, the approach can hardly be criticized. This posture, however, can now be re-evaluated in light of the drastic reduction in hardware costs.

An immediate corollary to the 'specialized data on selected patients' concept is that rigid 'standard nomenclature' must be devised, and enforced! The round patient is usually trimmed to fit the square computer. Although these generalizations are mandatory for a DDB approach, clarity may sometimes be sacrificed for conformity. A number of generalized systems: CPT, CRVS, ICD, ICDA, H-ICDA, ICHPPC, RCPG, SNOMED, SNOP, OXMIS; and a variety of local codes have been proposed. (Ref. 18). The very existence of so many competing schemes argues against the universal suitability of any one of them. Microprocessors, however, would work for smaller groups of doctors and, as such, have, at least, the possibility of providing a nomenclature that serves their expressed purpose, rather than an arbitrary system devised on some 'universal' scale.

With the above considerations in mind, what information would be necessary for each 'clinical entry', the basic unit of a patient oriented ADB? First, and foremost, the patient must be consistently identified. It is exceedingly important for security and safety, (recognizing the inviolate nature of Murphy's Law), that no assumptions be made as to whose data is whose. Some patient label must travel with each clinical entry. Second, the date and time should be marked. Third, and identification of the specific data type is made. The actual data may then be provided. In practice, though, certain embellishments of the above scheme should be entertained. Each of the four aspects needs to be discussed briefly.

Identification of the patients is most easily done with a 'computer number' that relates that patient to all other files. Hospital Numbers, Social Security numbers, or any other seemingly meaningless collection of random alphanumerica data are often awkward. Identification of time is best provided by storing the number of days elapsed from some reference date; ages and durations can then be derived. In addition, the day should be divided into convenient intervals so that a time can be associated with each clinical entry.

The identification of the general data types is the key to any patient oriented ADB system. The following have been employed: PROBLEMS, DIAGNOSES, FINDINGS, NUMERICALS (LABS), TREATMENTS, OPERATIONS, and MEDICATIONS. Certain routine administrative chores can be accomplished by adding EVENT/DATE and PAYMENT options to the above seven major categories. Finally, as a luxury, a FREETEXT option should probably be made available.

Having three categories, PROBLEMS, DIAGNOSES, and FINDINGS, may seem redundant, especially to a non-clinical analyst. The distinctions between them are often subtle, but not contrived. Take, for example, the term 'cyanosis'. This is most properly a PROBLEM, where 'pulmonary shunting' may be the FINDING on cardiac catheterization, and 'adult respiratory distress syndrome' is the final DIAGNOSIS. Patients may arrive with PROBLEMS that never acquire definitive DIAGNOSES, such as 'back pain'. Patients may have a FINDING, such as 'cardiomegaly', on chest X-ray, but that certainly can not be considered a DIAGNOSIS in the same sense as 'rheumatic mitral regurgitation'. If this distinction is preserved from the outset, all data will be specific. Given specific data, the computer can always generalize, but it can never do the reverse.

PROBLEMS, in the current approach, are associated with a time of onset and resolution. DIAGNOSES, on the other hand, are incorporated with two optional modifiers - (ie. acute, chronic, right, left, etc.). FINDINGS differ in that they are associated first with a modifier and then the method by which the FINDING was made. 'Cardiomegaly', for example, can be found by a number of different methods: physical examination, chest X-ray, echocardiogram, EKG, cardiac catheterization, operative exploration, or on post mortem. Each method carries different implications as to the confidence that can be placed in that particular finding. The ability to test the performance of the various diagnostic methods can only be provided if the method by which the finding was made is tabulated.

NUMERICALS contain data that is best characterized by numbers, rather than words. This choice is not always obvious, though, and depends primarily on the needs and preferences of the physician. Unless the clinician is specifically interested in the number, a FINDING of 'anemia' may be more useful than a 'hematocrit' of 30.6%.

TREATMENTS and OPERATIONS, again, overlap considerably, but the distinction is important. Performance of a tracheostomy or insertion of a thoracostomy tube may be considered either as an OPERATION or a TREATMENT, depending on the circumstances or requirements. The responsibility for that distinction, though, should not be placed on the programmer, but should, rather, be entrusted to the physician in charge.

PATIENT DATA/SCRATCH FORMATS

PAT=PATIENT NUMBER/P=PROBLEM NUMBER/D=DATA TYPE
T=TIME/R=REGULATOR/E=EXTRA

DATA TYPE	CODE ──────┐	CODE OF PROBLEM

PROBLEM   PAT | P | 1 | DATE | T ⇢ | DATE | T | R | E
RESOLUTION

DIAGNOSES   PAT | P | 2 | DATE | T ⇢ | MODE | MOD | R | E
CODE OF DIAGNOSIS

CODE OF FINDING
FINDINGS   PAT | P | 3 | DATE | T ⇢ | MOD | METH | R | E

CODE OF LABOROTORY TEST
LABORATORY   PAT | P | 4 | DATE | T ⇢ | VALUE | . | R | E

CODE OF TREATMENT
TREATMENT   PAT | P | 5 | DATE | T ⇢ | PRICE | . | R | E

CODE OF OPERATION
OPERATION   PAT | P | 6 | DATE | T ⇢ | PRICE | . | R | E

CODE OF MEDICATION
MEDICATION   PAT | P | 7 | DATE | T ⇢ | ? | ? | ? | ? | R | E
DISPENSE
DIRECTIONS   ±SOS
ROUTE   ±SUBSTITUE
SCHEDULE   ±PRN

INSTITUTION CODE
DATE   PAT | P | 8 | DATE | T ⇢ | X | X | X | X | R | E

PAYMENT   PAT | P | 9 | DATE | T | X X | PAYMENT | . | R | E

PAGE OF FREETEXT
FREETEXT   PAT | P | 10 | DATE | T ⇢ | X | X | X | X | R | E

FIGURE ONE

BYTE	1	2	3	4	5	6	7	8	9	10	11	12	13	14	15	16

MEDICATIONS, along with doses, schedules, and directions need little clarification. EVENT/DATE is used to flag processes not properly found in any of the other categories. These would include the date of admission, office visit, expiration, etc. Finally, if a critical piece of information will not gracefully fit into any of the above data types, it can be entered as FREETEXT in a separate file.

The actual nomenclature of the diagnoses, findings, etc., can not, of course, be efficiently retained in each clinical entry, but, instead, their positions on predefined internal translation lists are used. With the above specifications implemented on an 8 bit microprocessor, each clinical entry can be completely identified with a fifteen byte string. Because of the 128 byte/page format present in most standard random access files, an extra flag is added for convenience to each clinical entry so that the string is now sixteen characters long. The actual formats for each data type are shown in figure one. The complete clinical outline of most patients in the Division of Thoracic Surgery are contained in less than 400 sixteen bit bytes. This compares favorably with existing systems that require from 300-100,000 bytes per patient in storage. (Ref. 9).

There is considerable debate over the utility of certain record formats. Weed, (Ref. 16), feels that a Problem Oriented System maximizes patient care and has developed a method, PROMIS, in line with his beliefs. Others, (Ref. 7), consider that a Time-Oriented System best suits their clinical needs, and ARAMIS has been created with this in mind. The current system attempts to synthesize from the advantages of both and produces an integrated, chronological summary of the patient's problems.

Economy of storage and flexibility in information presentation, however, are not the major advantage of the system. The main utility of an ADB method resides in providing the user with extensive capabilities in data manipulation. One of the fundamental benefits of this approach is that none of the potential uses need be defined at the outset. The basic techniques – rearrangement of the clinical strings, searching, and sorting – are accomplished independent of the specific type of data. The original clinical data string in the Main Patient Data File is shown in figure 2. Sorting of the clinical strings arranged in the primary sequence, (1-2-3-4-5-6-7-8), will produce a series of problem oriented, chronological charts. Adjusting the string sequence and resorting can produce any number of useful results. For example, if the data type were shifted to position one, 'data code' to position two, and 'data/time' to position three, the new sequence, (3-5-4-1-2-6-7-8), Figure 2, would produce a file grouping all operations, treatments, diagnoses, etc., in chronologic order.

P R I M A R Y   S E Q U E N C E
Of Clinical Entry

1	2	3	4	5	6	7/8
PATIENT NUMBER	PROBLEM NUMBER	DATA TYPE	DATE/ TIME	DATA CODE	NUMERICAL VALUES	REGULATOR/ EXTRA

O P T I O N A L   S E Q U E N C E

1 (3)	2 (5)	3 (4)	4 (1)	5 (2)	6	7/8
DATA TYPE	DATA CODE	DATE/ TIME	PATIENT NUMBER	PROBLEM NUMBER	NUMERICAL VALUES	REGULATOR/ EXTRA

FIGURE TWO

NON-ALGORITHMIC DIAGNOSES (NAD)

Algorithmic diagnoses are usually based on Boolean strings, or their equivalent, that represent the weighted conditions considered necessary to establish a diagnosis. (Ref. 2,10,11). This approach has certain definite advantages. It is exact, precise, and rapid. One need merely define the disease, specify the population, weigh the criteria, and then plug in the patient values. Theoretically, this is quite attractive, but in reality, it possesses considerable drawbacks. The first is in the definition. It will be virtually impossible to reach a consensus regarding the definition of criteria that are sufficient and necessary to specify a disease. Very few authorities, indeed, can agree on precisely what it takes to establish a diagnosis. Certain results of this 'diagnosis by committee' could be rather expensive and somewhat dangerous for the patient. This point should not be considered as a mere technicality. Consider, for example, what is necessary to establish the diagnosis of pulmonary embolism. Some authorities would heparinize on the basis of clinical examination alone. Others would order blood gases, and a chest X-ray, despite the known limitaitons of these tests. Still others would require the security of scinti-scans and angiograms before they would consider instituting such potentially hazardous therapy. The more criteria that are employed, the greater the expense and danger to which the patient is subjected. In most instances, increasing the quantity of data usually sharpens the specificity of diagnosis; but at some point, though, diminishing returns will eventually be encountered. Obtaining a quorum on the determination of that point may be a formidable venture, at best. (Ref. 19).

Another criticism of algorithmic diagnosis is that it may fail to adjust to changing patient populations, or changing diagnostic capabilities. A new 'committee of experts' must be convened to establish further diagnostic criteria each time a new test is developed. The committee must also be prepared to establish the new weighting coefficients each time the population changes. How, then, would a Non-Algorithmic Diagnostic system address these problems?

An NAD relies solely on the patient population occupyuing its data base. It uses all findings and values as they are recorded, rather than some arbitrary, pre-defined subset. In taking advantage of the complete data pool, it automatically weights each test relative to the known population. Weaknesses in this type of NAD should be readily apparent, and they must by fully understood in order to prevent the placing of inappropriate confidence in the 'computerized' result. An NAD, by definition, has no source of information other than the clinical chart. This produces a very distinct population selection, which may enhance or detract from the derived result. In this sense, 'automatic weighting coefficient' and 'sample bias' are indistinguishable, but are certainly not interchangeable.

An NAD can not entertain a new diagnosis; it can only provide a frequency distribution of the diagnoses previously entered. An NAD can not recognize a zebra. It is also completely unworkable until a sufficient data base has been established; this process may be somewhat time consuming. Before the benefits of such an approach are discussed, though, an example of how NAD could work is offered.

Suppose, for example, a sixty year old man was admitted with chest pain and a normal chest X-ray, requiring diagnostic work-up. The ADB files would first be searched for all males, over, perhaps, forty, who had chest pain and a normal study on chest X-ray. A frequency distribution of all diagnoses established on similar patients would then be constructed from the ADB. This may, for example, demonstrate that on one hundred patients with the same findings, forty-two had pulmonary emboli, and forty-seven and acute myocardial infarctions. The object, then, would be to discriminate between these two diseases. Frequency distributions of all findings and laboratory would then be separately produced for pulmonary emboli and myocaridal infarctions. These frequency distributions would then be contrasted to isolate those tests or findings which most consistently distinguish the disease entities. It may turn out, for example, that the white cell count was elevated in ninety five percent of the myocardial infarctions, but also in just as many patients with pulmonary emboli. Although the white count may be elevated in a high percentage of both subsets, it offers no more advantage in distinguishing these particular diagnoses than tests that are seldom positive. In other words, an NAD could direct the patient's evaluation in a statistically optimal fashion. After these 'optimal' tests are obtained, the files could be searched once again, but with added information. The new diagonsis frequency distribution may therefore be sharpened. Now perhaps pulmonary emboli may occur in eighty five percent of the patients, and myocardial infarctions in only ten percent. There is no theoretical limit to the number of times that these recursions could be performed. If adequate information is present in the ADB, progressively improved differential diagnoses could be continously formulated.

An NAD system then 'learns' medicine in much the same way as a physician: he is first exposed to a variety of clinical situations, then attempts to remember the salient features of the disease. His ability to do this is often uncanny. An NAD system does precisely the same thing. Because no prior definitiions of the disease have been created, the NAD does not labor under any preconceived limitations. Provided the data is available and accurate, no information is exempt from analysis, and nothing is forgotten. An NAD is a plastic system; it would not necessarily formulate the same differential diagnosis from one month to the next. It can spot trends where an algorithmic system can not. By its very nature, though, it is dependent on selection, and generalizations can only be made relative to the sampled data base. This is an extremely important consideration, and must constantly be kept in mind to prevent the uninitiated from placing unrealistic confidence in a 'computer formulated' diagnosis.

SUMMARY

As in any field of endeavor, most things are bought, and little is free. In data base management, speed is bought with flexibility, convenience with complexity, and space with time. In order for a system to be forgiving, it must have been imparted considerable understanding. Microprocessors have extended the invitation of automated data management to an enormous segment of the medical community; an invitation usually reserved for the financially privileged. As such, many physicians must now make the decisions - and mistakes - that their mainframe forerunners confronted years ago; but because of the limited number of patients they may now directly manage, considerable greater detail may be incorporated into the patients file, and the ADB is one method by which this may be accomplished. In addition, the NAD is a logical extentsion of the ADB, and if its limitations are recognized, it could become a powerful tool in improved patient care.

SYSTEM

ALTAIR 8800b Microprocessor
64 K RAM Memory
ALTAIR Hard Disk Controller
ALTAIR 10 Megabyte Disk Drive
Microsoft Extended Disk Basic, Version 5.0, 1979
APOLLO Patient Management System, Brandon Systems
Hazeltine 15000 CRT; QUME Q70 Printer

REFERENCES

1. BARNETT, G.O., ET AL.: 'MUMPS: A Support for Medical Information Systems', Medical Informatics, (1), pp. 183-189, 1976.

2. BLEICH, H.L.,: 'The Computer as a Consultant', New England Journal of Medicine, (284), pp. 141-147, 1971

3. BRADSHAW-SMITH, J.H.: 'A computer Record Keeping System for General Practice', British Journal of Medicine, (1), pp. 1395-1397, 1976.

4. CARROLL, D.F.: 'Arizona Health Plan Data Management System', in Zimmerman (1974), pp. 26-32, 1974.

5. COE, F.L.: 'The performance of a Computer System for Metabolic Assessment of Patients with Nephrolithiasis', Computers in Biomedical Research, (7), pp. 40-55, 1974.

6. COLLEN, M.F.: 'Automated Multiphasic Health Testing: Implementation of a System', Hospitals, (45), pp. 49-57, 1971.

7. FRIES, J.F.: 'Time-Oriented Patient Records and a Computer Data-bank', Journal of the American Medical Association, (222), pp. 1536-1542, 1972.

8. FRIES, J.F.: "ARAMIS: American Rheumatism Association Medical Information System', Second Year Progress Report under Grant HS-1875 from the National Center for Health Services Research, 1977.

9. HENLEY, R.R., WIEDERHOLD, G.: 'An Analysis of Automated Ambulatory Medical Medical Record Systems', (PB-254-234), National Technical Information Service, US Department of Commerce, 5285 Port Royal Road, Springfield, Va., 1975.

10. HOLLAND, R.R.: 'Decision Tables. Their Use for the Presentation of Clinical Algorithm', Journal of the American Medical Association, 33), pp. 455-457, 1975.

11. LEDLEY, R. S., LUSTED, L.B., 'Reasoning Foundations of Medical Diagnosis', Science, (130) pp. 9-21, 1959.

12. McDONALD, C.J., ET AL: 'A computer-Based Record and Clinical Monitoring System for Ambulatory Care', AMERICAN JOURNAL OF Public Health, (67), pp. 240-245, 1977.

13. PACE, G.C., BRANDEJS, J.F.: Principles of Computer-Aided Physician's Office (CAPO), the Canadian Medical Association, Ottawa, Canada, 1975.

14. PERRY, J.: ' Medical Information Systems in General Practice: A Community Health Project', Proceedings of the Royal Society of Medicine, (65), pp. 241-242, 1972.

15. SANAZARO, P.J., 'Automated Multiphasic Health Testing: Definition of the Concept', Hospitals, (45), pp. 41-43, 1971.

16. WEED, L.L., 'Medical Records that Guide and Teach', New England Journal of Medicine, (278), pp. 593-599, 1968.

17. ZIMMERMAN, J.,Editor, "Proceedings of the 1973 MUMPS Users' Group Meeting', Published by the MUMPS Users' Group, c/o Mr. Richard E. Zapolin, The MITRE Corporation, PO Box 208, Bedford, Ma., 1974.

18. ZIMMERMAN, J., RECTOR, A., Computers for the Physician's Office Research Studies Press, PO Box 2, Forest Grove, Oregon, pp. 124-130, 1978.

19. ZIMMERMAN, J., RECTOR, A., Computers for the Physician's Office Research Studies Press, PO Box 2, Forest Grove, Oregon, pp. 104-105, 1978.

20. ZIMMERMAN, J., RECTOR, A., Computers for the Physician's Office Research Studies Press, PO Box 2, Forest Grove, Oregon, pp. 35-74, 1978.

CRITICAL CARE INSTRUMENTATION: AN APPROACH USING MODULAR ELEMENTS AND DISTRIBUTED PROCESSING

R.W. Hagen, M.S., M.W., Browder, M.S.,
W.R. Roloff, B.S., L.J. Thomas, Jr., M.D.

Biomedical Computer Laboratory
Washington University
St. Louis, Missouri

INTRODUCTION

The changing needs that characterize critical care instrumentation systems are not unique. This dynamic situation prevails in most active research environments where biomedical instrumentation is employed. Our experience indicates that these systems evolve in such a way that new needs are determined by the outcomes of current experiments making it impossible to predict long- range requirements satisfactorily. In the past, a degree of flexibility was gained through the introduction of minicomputers into medical research activities(1,2). Since the beginning of the "microprocessor revolution", the microcomputer has been applied either as a less expensive and less powerful substitute for an instrumentation system's centralized minicomputer or as an inaccessible component imbedded within a particular instrument. Through experience in applying microcomputers, another role for this new tool has emerged(3). The CPRS concept embodies this new role in that it depends on a distributed set of programmable microcomputers packaged in function-specific modules which communicate over a standard digital bus. Through this approach, the implementation of biomedical instrumentation systems can be effectively addressed.

The scheme adopted for the CPRS is general enough to enable the addition of both experimental and established measurements to previously configured systems. While the required transduction techniques and control devices may differ from application to application, many critical care applications have common underlying

needs for signal acquisition, feature extraction, information display, information storage and control(4). The processing power
that can be economically focused on these common needs continues to
increase as the offerings of electronic component manufactures improve almost daily(5). However, the low-cost virtue of microprocessors cannot be realized until a substantial investment has been
made in support equipment, support personnel and an implmentation
procedure.

Several commercial systems, particularly suited for use in
scientific laboratories, have recognized some commonality in underlying needs. Such systems are marketed by DEC and IBM. The
MINC-11 system offered by Digital Equipment Corporation features a
centralized microcomputer interfaced to dual floppy discs, a dis-
play/keyboard terminal, the IEEE-488 bus, and MINC modules. Each
MINC module provides a specific function such as A-to-D conversion
or D-to-A conversion. A similar product is IBM's Device Coupler.
It features an ensemble of modular I/O units containing built-in
functions. The Device Coupler may act as a terminal to a host computer system. The modular I/O functions include a digital-input
module, a digital-output module, an analog input module, and a
scope-plot module. IBM reports that in a survey of over 200 applications at the T. J. Watson Research Center, more than 70 percent
of the applications surveyed fell into the requirement category for
which the Device Coupler was a viable tool(6).

SYSTEM DESCRIPTION

The CPRS design focuses on a modular approach which spans
hardware, software and packaging allowing instrumentation systems
to be configured appropriately for each particular application, yet
maintaining the flexibility necessary to change with needs. Each
hardware module is tailored to provide a specific function.
Although each module is functionally different and operates inde-
pendently in time, they are coupled to one another through a standard communications pathway. Because the functional portion of
each module is isolated from the common pathway, the design of new
modules is not constrained, allowing the designer of new modules to
take advantage of current hardware and software offerings, while
remaining compatible with previously designed modules. Software
modularity is achieved through the use of a
microprocessor-development system which contains a growing library
of subprograms written in assembly language. The application program for a particular functional module is constructed by linking
together the appropriate subprograms using the development system.

This microprocessor-development system aids in entering, as-
sembling, and loading programs along with monitoring their opera-
tion(7). Program entry and assembly are done on the host computer

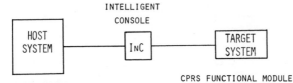

INTELLIGENT
CONSOLE

HOST
SYSTEM

InC

TARGET
SYSTEM

CPRS FUNCTIONAL MODULE

HOST PROCESSOR
OPERATING SYSTEM
MASS STORAGE
CTR/KBD TERMINAL
LINE PRINTER

Figure 1.   Microprocessor Development System

system which has facilities to edit, store, and retrieve symbolic
programs.   Target-microprocessor programs, in symbolic form, are
manipulated by using the host system's editor until the program is
ready for assembly.   The symbolic program is then translated to
target-object code by a cross-assembler which executes on the host
computer.   Once a target program is loaded into the target system
it is necessary to control and monitor its operation.   In the case
of the CPRS, the target is a functional module. The intelligent
console, InC, provides for monitoring and control of the target microprocessor system as the target system executes the program.
After a program is fully debugged the intelligent console is removed.

IMPLEMENTATION

In meeting the CPRS goal of providing a practical and systematic approach to the application of microcomputers to medical instrumentation tasks, three important issues were identified:

   * the proper distribution of processing tasks

   * the choice of a standard intermodule communications pathway

   * the selection of a packaging scheme

Since the distribution of tasks has a direct impact on overall
system performance, the point of separation of these tasks must be
carefully chosen.   The CPRS gives emphasis to design segmentation,
requiring the identification of tasks that can be performed inde-
pendently and asynchronously. Currently we have implemented four
types of CPRS modules: a user interface module, a signal processing module, a data storage module, and a controller module. Each
module contains a kernel microcomputer and a few function specific
components such as an analog-to-digital converter in the signal
processing module.   In addition to these modules, any device compatible with the standard communciations pathway can be included in
an instrumentation system.

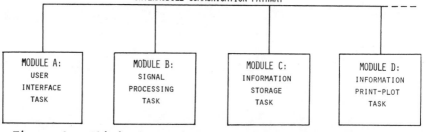

Figure 2.   Clinical Physiological Research System Diagram.

The second important issue is the choice of the intermodule communication pathway. Keeping in mind that distributed systems can be limited by the capability of this interconnect, that a module will view other modules only through the interconnect and that the proper choice of an interconnect will make many commercial devices available, several candidate bus structures were compared with projected CPRS requirements. Technical specifications such as data rate, bus length and number of devices were considered. Non-technical considerations included cost effectiveness, ease of use, and anticipated acceptance of the bus standard. The IEEE Standard-488 Digital Interface for Programmable Instrumentation was chosen because its characteristics best suited CPRS objectives.

An important but often neglected issue in developing a practical method for implementing instrumentation systems for critical care applications is packaging. Cost is of course important, but we were also interested in a packaging scheme which would allow us to reconfigure systems as an application requires, to construct portable systems which were suitable for use in a patient's room, and to expedite module construction. The TM500 series of packaging components, manufactured by Tektronics was chosen. A wire wrap panel with on-board power regulation was designed to substitute for the standard circuit board supplied with each plug-in kit. Functional modules are constructed in the kits and inserted into the TM500 mainframe which is modified to accomodate the 488 bus.

The figure shows signal processor, user interface and bus controller modules in a mainframe. Unregulated power is distributed to each module connector. The mainframe contains a rear-mounted IEEE-488 standard connector. The printer/plotter shown will generate any one of the standard ASCII characters by sending the appropriate code and location to the plotter over the 488 bus. The plotter carries out the linear interpolation necessary to connect two points with a straight line. This printer/plotter serves to illustrate that cost-effective peripheral devices are increasingly able to perform tasks which a centralized minicomputer formerly handled.

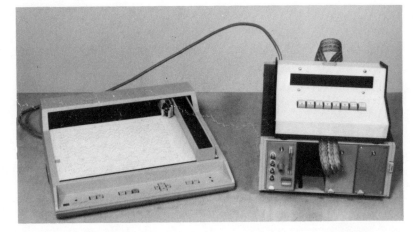

Figure 3.   Clinical Physiological Research System Modules.

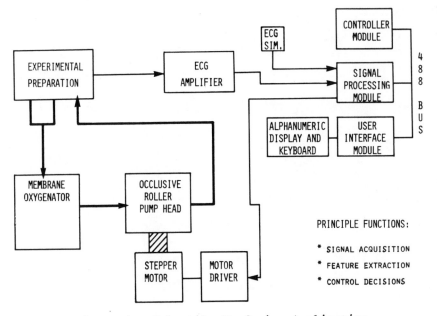

Figure 4.   Pulsatile Perfusion Application

APPLICATIONS

Several systems have been configured using the CPRS.   One of
these systems is used to study physiologic responses to changes in
pulsatile perfusion variables during extracorporeal perfusion for
anesthetized dogs.   In this system it is necessary to extract the

QRS feature from the electrocardiogram and control the arterial blood flow returning to the animal after the blood has been oxygenated. This controlled blod flow is a function of the animals heart rate, the QRS occurrence, and three perfusion variables entered by the user.

The same QRS-detection algorithm is employed in a patient monitoring application. The modular QRS-detection subprogram was beneficial in this situation. This patient-monitoring system uses the same set of CPRS hardware modules as the pulsatile perfusion system. A commercially available printer/plotter is used to display processed physiologic information for patients in a critical care environment.

Our objective to develop a systematic and practical approach to applying microprocessors to the task of implementing a significant class of biomedical instrumentation systems can be realized through the CPRS. The Clinical Physiologic Research System has enhanced our ability to respond to changing instrumentation-system requirements in laboratories and in the patient-care environment.

This work was supported by research grant RR-00396 from the Division of Research Resources of the National Institutes of Health, Bethesda, Maryland.

REFERENCES

1. HAGEN, RW, THOMAS, JR., LJ, AND JOHNSON, JA: "Design Consider-
        ations for a Computer-Based Clinical Physiological Re-
        search System," Proceedings of the Annual Association for
        Computing Machinery Conference, Houston, Texas, pp.
        99-104, October 20-22, 1976.

2. DEWEY, JR., CF, METZINGER, RW, AND LEES,, RS: "A Small Repli-
        cable Computer System for Clinical Analysis," Proceedings
        of the San Diego Biomedical Symposium, vol. 15, 1976.

3. SHIPTON, HW: "The Microprocessor, A New Tool for the Biosci-
        ences," Annual Review of Biophysics and Bioengineering,
        vol. 8, pp. 269-286, 1979.

4. GLAESER, DH, AND THOMAS, JR., LJ: "Computer Monitoring in Pa-
        tient Care," Annual Review of Biophysics and
        Bioengineering, vol. 4, pp. 449-476, 1975.

5. QUEYSSAC, D: "Projecting VLSI's Impact on Microprocessors,"
        IEEE Spectrum, vol. 16, no. 5, pp. 38-41, May 1979.

6. GUIDO, AA, COLE, H, AND KREIGHBAUM, LB: "Interactive Laborato-
        ry Automation Aid: The Research Device Coupler,"
        Proceedings of the IEEE, vol. 63, no. 10, pp.
        1509-1513, October 1975.

7. SPENNER, BF, BLAINE, GJ, BROWDER, MW, HARTZ, RK, AND JOHNS, GC:
        "Microprocessor-Based Microprocessor Instrumentation: An
        Intelligent Console," Proceedings of the National Bureau
        of Standards Conference on Microprocessor Based
        Instrumentation, Gaithersburg, Maryland, pp. 1-5, June
        1978.

KNOWLEDGE ENGINEERING IN CLINICAL CARE

J.C. Kunz*, L.M. Fagen°,
E.A. Feigenbaum°, J.J. Osborn*

°Computer Science Department
Stanford University
Stanford Ca. 94305

*The Institutes of Medical Sciences
2200 Webster St.
San Francisco, Ca. 94115

During the care of critically ill patients, pertinent data may be obscured by the large amount of information in the patient record. we have developed a computer program to interpret measured physiological data, recognize the clinical context in which the data were measured and in which they must be interpreted, extract important data, and bring interpretations to the attention of the bedside clinician. The program uses a set of rules which contain "knowledge" about critical care patient management and "expectations" about the future possible interpretation of the patient physiology as it changes in time. The rules are statements of the expert clinical practitioner about the practice of critical care medicine. Each rule is of the form "if a set of facts are true," then "make a conclusion based on these facts". The rules are written in an English-like syntax using clinical terminology. Use of rules for representation and manipulation of knowledge about a problem is taken from the subdiscipline of Artificial Intelligence called "Knowledge Engineering".

We are developing a rule base for a specific well-defined problem, ventilator management. Measured data are entered into the program interprets and reports deviations from expectations and readiness of the patient for new therapeutic goals. The system is intended to complement normal staff activity. The focus of the system is on clinical decision making: interpreting the clinical significance of measured data and relating the implications of measured data to patient physiological state and therapeutic goals for patient management. We expect the automated decision assistance to be particularly important for recognizing infrequently occurring but important clinical events, ie recognizing the situations that human observers do not consistently recognize in a timely manner.

The Ventilator Manager program (VM) interprets the clinical significance of quantitative data used in managing patients receiving ventilatory assistance in the intensive care unit (ICU). An extension of a physiological monitoring system VM (1) provides a summary of the patient physiological status appropriate for the clinician; (2) recognizes untoward events in the patient/machine system and provides suggestions for corrective action; (3) suggests adjustments to ventilatory therapy based on long-term assessment of the patient status and therapeutic goals; (4) detects possible measurement errors; and, (5) maintains a set of patient specific expectiations and goals by which data are evaluated.

The program produces interpretations of the time-varying physiological measurements using a model of the therapeutic procedures in the ICU and clinical knowledge about the diagnostic implications of the data. These clinical guidelines are represented by a knowledge-base of the rules specified by expert users of ICU data.

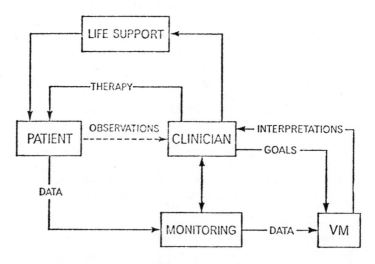

Figure 1. VM System. Physiological Monitoring is now often added to the classical patient-clinician relation in the ICU. VM interpretation is an extension to computer-based patient monitoring.

CLINICAL APPROACH TO ICU MONITORING DATA INTERPRETATION

The VM knowledge base is conceived upon the concept that diagnosis should be made, to the extent possible, using physiological data and reasoning which attempts to infer the physiological state of the patient given relevant quantitative data.

The clinical approach to interpretaion of ICU patient data includes several different phases. These phases are incoporated into the VM clinical knowledge base as levels of abstraction. The VM knowledge base includes procedures by which the system will:

1. Characterize measured data as reasonable or spurious;

2. Determine therapeutic state of the patient (currently the mode of ventilation);

3. Define the expected physiologic meaning of future values of measured variables;

4. Check physiologic status, including: cardiac rate, hemodynamics, ventilation, oxygenation;

5. Identify appropriate therapy, including management of abnormal measurements and readiness of the patient for therapies which are consistent with general therapeutic goals.

The system currently has only skeletal knowledge about each of these basic kinds of interpretation.

In order not to misinterpret the clinical situation, VM will be provided with rules to recognize problems in the ICU which cause spurious data, including factors such as staff actions which invalidate measurements (eg: disconnecting an arterial line to draw a blood sample, thereby invalidating blood pressure measurements), routine problems with measurement devices (eg: clogging of airway sampling lines), and unusual hardware failures (eg: occlusion of the ventilator expiratory valve). The system has knowledge about the relation between failures and characteristic abnormalities in measured data; and it can identify one or a small number of alternative causes of spurious data based on simultaneous analysis of multiple data sources.

The system has knowledge about the states of ventilatory therapy, including definitions of and characteristics of spontaneous breathing, continuous mandatory ventilation, assisted ventilation, and T-piece breathing. VM will directly use ventilator status when a ventilator becomes available which provides direct computer sensing of its operating mode. Until that time, ventilation mode is inferred by analyzing characteristic changes in the times of inspiration and exhalation and the maximum inspiratory pressures. The system now determines expected values for measured parameters based upon the current therapeutic state of the patient.

Diagnoses are made in terms of expected measurement values for a patient in a particular therapeutic situation. For example, the definition of hypoventilation is made not in terms of a fixed pCO_2, but rather upon the range of pCO_2 which is reasonable and appropriate for the particular clinical state of the patient. Factors considered in determining appropriate ranges include ventilatory therapy (eg mandatory ventilation, T-piece), and prior history of respiratory disease. Heart rate, etc., are interpreted in a similar fashion, depending upon the current clinical therapy and the immediate and long-term patient history.

System Organization

VM has three basic parts: the knowledge base, an interpretation program which applies the patient data to the knowledge base, and a parser which converts rules in the knowledge base into a format usable by the interpreter. The knowledge base, consisting of a set of production rules, uses clinical terminology and a stylized English-like syntax. The interpretation program has a number of functions including: input of clinical data and output of messages to the user and to the monitoring system computer; specification of the operation of the functions defined in the knowledge base (eg stability of measurements over time); and an interpreter which invokes rules, determines if rules are applicable and true, and carries out the specified actions implied by the conclusions of rules found to be true. The parser converts rules from their English-like syntax to a syntax which can be directly interpreted by the rule interpreter. The interpreter and the parser are coded in the LISP symbolic processing language.

VM uses production rules to encode knowledge about ICU diagnosis and therapy. The rules in VM are of the form: IF Statements about measurements and previous conclusions are true THEN make a conclusion based on these facts; make appropriate suggestions to clinicians; and (possibly) create new expectations about the future acceptable ranges for measured variables. Additional information associated with each rule includes a comment about the main concept of the rule and a list of all of the therapeutic states in which it makes sense. Figure 2 shows a sample rule for determining hemodynamic stability.

STATUS RULE: STABLE-HEMODYNAMICS
DEFINITION: Defines stable hemodynamics for most settings
APPLIES to patients on VOLUME, CMV, ASSIST, T-PIECE
COMMENT: Look at mean arterial pressure for changes in
     blood pressure and systolic pressure for maximum
     pressures.

IF
  HEART RATE is ACCEPTABLE
  PULSE RATE does NOT CHANGE by 20 beats/minute in 15 minutes
  MEAN ARTERIAL PRESSURE is ACCEPTABLE
  MEAN ARTERIAL PRESSURE does NOT CHANGE by torr in 15 minutes
  SYSTOLIC BLOOD PRESSURE is ACCEPTABLE
THEN
  the HEMODYNAMICS are STABLE

Figure 2: Sample VM Interpretation Rule. The meaning
of "ACCEPTABLE" varies with the clinical context, ie
whether the patient is receiving VOLUME or CMV ventilation,
etc. This rule makes a conclusion for internal system use.
Similar rules also make suggestions to the user.

Most of the rules represent the measurement values symbolically, using the terms of "acceptable" or "ideal" ranges. The actual
meaning of "acceptable" is changed as the patient moves from state
to state, but the statement of the relations between the physiol-
oggical measurements remains constant. Use of symbolic statements
about physiology (eg. 'heart rate is acceptable') allows rules to
make uniform statements about physiology which will be applicable
in clinical situations. Use of symbolic statements in this way represent common clinical practice; it shows common principles of
physiological interpretation in different contexts; and it minim-
izes the number of rules needed to describe the complexity of the
diagnostic situation.

The meaning of the symbolic ranges is determined by rules that
establish the context dependent expectations about the value of
measured data. For example, when a patient is put on the T-piece,
the upper limit of acceptability for the expired pCO_2 is raised because minute volume normally decreases temporarily when the mechanical ventilator is disconnected.   An example rule that creates
these expectations is shown in Figure 3.

INITIALIZING RULE;   INITIALIZE-CMV
DEFINITION: Initialize expectations for patients on CMV
APPLIES to all patients on CMV
IF ONE OF
        PATIENT TRANSITIONED FROM VOLUME TO CMV
        PATIENT TRANSITIONED FROM ASSIST TO CMV
THEN EXPECT THE FOLLOWING

	very low	low	[- ideal -] min	[- ideal -] max	high	very high
	----	----	----	----	----	----
MEAN ART. PRESSURE	+ 60	+ 75	+ 80	+ 95	+ 110	+ 120
HEART RATE	+	+ 60	+	+	+ 110	+
END TIDAL CO_2	+ 22	+ 28	+ 30	+ 35	+ 42	+ 50

. . .

Figure 3.   Portion of an Initializing Rule.   This
rule establishes initial expectations of acceptable
and ideal ranges of variables.

Physiological definitions, eg "Hemodynamic stability" are used
with other conditions, such as the absence of hyperventilation, to
identify appropriate therapy, eg. possible readiness of the patient who has been receiving assisted ventilation to be placed on the
T-piece.   Since many therapeutic decisions require bedside observation, the system makes qualified statements about appropriateness
of therapy.

COMPUTER BASED FLUID BALANCE RECORDS *+

John E. Brimm, M.D., Mark Halloran **,
Clifford M. Janson, Richard M. Peters, M.D.

Department of Surgery/Cardiothoracic
University of California Medical Center
San Diego

and

** Hewlett Packard Co.
Waltham, Massachusetts

* Presented at the First Annual International Symposium Computers in Critical care and Pulmonary Medicine, Norwalk, Connecticut, May 24-26, 1979.

+ Supported by USPHS Grant No. GM 17284 and ONR Contract N00014-76-C-0282

Fluid balance data is essential for decision making regarding cardiovascular, respiratory, renal, and metabolic function; however, manual charting of fluid balance is characterized by three principal deficiencies: it is inaccurate, it reflects fluid volumes but not their compositions; and it masks long-term effects. In order to assess the accuracy of manual charting, we studied the records of 30 randomly selected patients, whose ICU admissions lasted four or more days, from five 6-bed ICU's in our hospital. We found that 30 percent of the daily records were incorrect, and that only four patients had fully error-free records. Most errors were arithmetical; however, errors persisted even when calculators were used. Frequently, hourly entries were omitted from totals, and intakes were confused with outputs. In one case, a net intake of two liters was confused with a net output. These mistakes, as expected, were correlated with the number of intakes and outputs, indicating that patients with the most complex fluid management problems are least likely to have accurate records. The implications of this inaccuracy are clear. In clinical practice, physicians really cannot rely solely on fluid balance records for guiding therapy; rather, they utilize other clinical data, such as serum electrolyte measurements, for controlling fluid management. The economic implication of this over-reliance on laboratory measurements is enormous. A computer-based system, unlike manual charting, can assure systematic storage and accurate arithmetic. Consequently, it has the potential to reduce reliance on the clinical laboratory.

Because manual charting is both time consuming and inaccurate, it has typically been used only for tallies of fluid volumes. Thus, despite the fact that all intakes have known composition and that the composition of outputs can be estimated or measured, manual charting does not provide information on electrolyte balance or nutritional intake. An automated system can provide a much more detailed and complete report of this information for fluid and electrolyte management.

A further limitation of manual charting is that the records typically only present information for the current day. Thus, at the beginning of the day, a physician reviewing the record must look through two different pages to determine recent events. Although information is accumulated during the nursing shift, running balances are rarely kept and are calculated only at the end of the shift. Because ICU flow sheets have been designed to highlight recently occurring events, they may mask effects occurring over longer periods. For the management of vital organ function in the ICU, most events occurring two days in the past are unimportant. In fluid management, however, past events, or the cumulative effect of past events, may be crucial. For example, inadequate provision for sodium losses may result in hyponatremia and systemic underreplacement of chloride for a patient on nasogastric suction may manifest itself in metabolic alkalosis. These events may evolve over four to five days. Each of these time effects − discontinuities at page boundaries, the lack of running balances, and the masking of long-term effects − can be circumvented through the use of electronic records.

Because of the recognized deficiencies of manual charting of fluid balance, computer-based records offer tremendous potential. In 1969, Warner(1) reported a system for keeping volume balances. Reichertz(2) and Kloevekorn et al.(3) have developed systems which utilize fluid compositions for calculating eletrolyte and nutritional balances. Siegel (4,5) has described a much more complex system, based on the work of Moore,(6) that keeps sodium, potassium, chloride, and nitrogen balances and uses them for estimating patients' fluid, blood, electrolyte, and nutritional requirements.

We began development of a computer-based system for keeping intake/ output (I/0) records about four years ago. Since the I/0

programs were intended for use in a commercially available monitoring system, the Hewlett-Packard 5600A Patient Data Management System, the underlying design had to provide features permitting
transfer to other institutions for use in caring for both adult and
pediatric patients with many different acute clinical problems. In
a previous report, (7) we described some of the problems and our
design solutions for making the system adaptable and acceptable to
nurses and physicians not participating directly in its development.

PROBLEMS OF TRANSFERENCE

The problems of developing computer systems capable of being
used in many settings are enormous, and consequently have impeded
the acceptance and impact of medical computing applications.
Typically the problems arise from incompatible hardware and
software, but, in the case of our I/O system, all potential users
had basically the same hardware and software. The difficulty in
developing a transportable system arose from fundamental user
differences. We wanted the system to be acceptable to physicians
and nurses with different backgrounds, experience, and specialties.
Despite agreement on the medical rationale for the system, potential users differed significantly on its proposed operational details. These differences can be classified as:

a) Institutional: The system had to provide for the different
IV solutions and blood products used in different hospitals or in
different ICUs in the same hospital. As a trivial example of the
problems encountered, we had to accommodate the nursing schedules
used in different hospitals. The start of the "day" and nursing
shifts had to be configurable for accumulating fluid totals. In
our hospital, a day starts at 6:01 a.m. one day and runs until
6:00 a.m. the next; in other hospitals, other schedules are used.

b) Medical: The system was intended for use with different
patient populations - neonatal, pediatric, trauma, burn, respiratory, and cardiac surgical. Considerations important in one setting
might be relatively unimportant in another; consequently, the system had to be capable of being tailored to the particular clinical
setting.

c) Idiosyncratic: Many differences suggested by potential
users were attributable to human foibles, whims, or preferences.
Some users, for example, prefer horizontally formatted reports
while others like a vertical format. Others want the order of
items on a report changed. These differences rarely affected the
basic medical content or underlying system structure, but they may
be time consuming to accommodate. Provision for these differences
may be crucial for achieving acceptance.

Design of the system to provide for these institutional, medical, and idiosyncratic differences has been essential for making the system adaptable to various settings. Moreover, because medical requirements and personnel have changed in our hospital, the system has had to evolve even within our setting. Recently, for example, physicians were interested in tallying nitrogen balances and the amount of albumin infused. The IV solutions used are constantly changing. These requirements have been easily met. Thus, the provision of features permitting transference of the system to other settings has fostered easy evolution as well.

SYSTEM DESIGN

The system is structured as a series of user-configurable tables. Initially, users must determine the constituents that they want the system to tally. Most will choose a default set including electrolytes - sodium, potassium, chloride - and nutritional elements - protein, carbohydrates, and calories. Potential users could not agree on all components, however. Some wanted to distinguish orally given calories from intravenous; other wanted to keep nitrogen balances; still others wanted to tally certain drugs. By making these constituents user-definable, the system can be customized to their requirements. As mentioned, this capability permits evolution without reprogramming as medical practice changes.

```
                          FLUID   LIST

                     17 PROT 8.5%  ORAL
     IV'S            18 PROT 5%
                     19 HYPR SALIN 29 ORAL FLUID
      1 D5/W                       30 MEATBASE
      2 D5 1/2 NS                  31 ENSURE
      3 D5 1/4 NS                  32 ISOCAL
      4 D10/W         BLD PRODUCTS 33 VIVONEX-HN
      5 D50/W                      34 MYLANTA
      6 NS           20 WHOL BLOOD 35 RIOPAN
      7 1/2 NS       21 PLASMA     36 AMPHOGEL
      8 RL           22 FRZ PLASMA 37 MAALOX
      9 D5 NS        23 PLASMANATE
     10 D5 RL        24 FRZ R CELL
     11 NAHCO3 8.4   25 PACKD CELL
     12 D5 ISLYT M   26 PLATELETS  OUTPUTS
     13 MANNITOL     27 ALBUMIN25%
     14 INTRALIPID   28 ALBUMIN 5% 38 BLOOD OUT
     15 FREAMINE 2                 39 BLOOD-TEST
     16 KCL (MEQ)                  40 URINE

     FOR INSTRUCTIONS PUSH '; GO'
     ENTRY :1;7;8;
```

Figure 1. The list of intakes and outputs currently configured for our hospital.

```
    111111111    TESTING              20 JUN 1127

   FREAMINE 2 +ADD#  1  - ADDITIVE SELECTION

   ADDITIVE        |AMT|U/L|  |COMP| STD| TOT|U/L
      1 D50/W      |500|MLS|  |VOL | 500|1000|MLS
      2 D10/W      |  0|MLS|  |NA  |  10|  55|MEQ
      3 KCL        | 30|MEQ|  |K   |   0|  30|MEQ
      4 NACL       | 50|MEQ|  |CA  |   0|   0|MEQ
   -> KH2PO4       |  0|MEQ|  |MG  |   0|   0|MEQ
      6 CA GLUCON  |  0|MEQ|  |CL  |   0|  80|MEQ
      7 K ACETATE  |  0|MEQ|  |HCO3|   0|   0|MEQ
      8 MGSO4      |  0|MEQ|  |HPO4|  20|  10|MEQ
      9 CACL2      |  0|MEQ|  |SO4 |   0|   0|MEQ
     10 NA         |  0|MEQ|  |BASE|   0|   0|MEQ
     11 K          |  0|MEQ|  |HGB |   0|   0|GM
     12 CA         |  0|MEQ|  |CAL | 312|1006|CAL
     13 MG         |  0|MEQ|  |PROT|  78|  39|GM
     14 CL         |  0|MEQ|  |NITR|  13|   6|GM
     15 HPO4       |  0|MEQ|  |CARB|   0| 250|GM
     99 '-. GO' TO STORE      |FATS|   0|   0|GM
   FOR INSTRUCTIONS PUSH '; GO'
   ENTRY :                           PAGE   1 OF  1
```

Figure 2. Display used for specifying hyperalimentation solutions. Explanation of this display is given in the text.

Once users have configured the table of constituents, they may then define a second table of the commonly occurring intakes, outputs, and additives for their settings. A listing of some of the fluids configured for out hospital is illustrated in Figure 1. Each of the fluids of the list is defined in terms of the concentration of each of the constituents configured in the first table. Thus, normal saline would be defined as 154 mEg/L of both sodium and chloride in a liter of water, with no protein, calories, etc. Each fluid is grouped into one of up to eight categories; in our implementation, we use only four: IV's, blood products, oral fluids, and outputs. Users may add or delete fluids at any time, a process which takes less than five minutes at a programmer's console.

While intakes have known compositions, outputs do not. We estimate the composition of some drainages, such as chest drainage, to have the same composition as serum. Other outputs such as urine, however, are so variable that they cannot be estimated accurately and must be measured. Consequently, we have provided a capability for entering the concentrations of sodium, potassium, chloride, etc., when they have been measured for outputs. A similar capability has been made for specifying the composition of highly variable intakes, such as hyperalimentation solutions and oral fluid dilutions. The composition of hyperalimentation solutions depends on a patient's requirements, and the possible combinations are too numerous to be listed exhaustively. Thus, they must be specified at the time that they are given, as shown in Figure 2. In this display, the standard solution is listed at the top, the possible additives are listed on the left margin, and the constituent are listed on the right. The composition of the standard is listed under "STD", and the composition of the resulting solution, standard plus additives, is listed under "TOT". The additives are user definable and may be salts, drugs, or IV solutions. The units of additives may be milliliters (MLS), milliequivalents (MEQ), ampules (AMP), or grams (GMS). The system automatically adjusts all constituent concentrations to common units. A user may specify the additives to a standard solution by moving the arrow to the desired line and entering the amount of that additive. As each entry is made, the concentrations under the "TOT" label are updated. In the example shown, the IV admixture is composed of each part (500 ml) of Freamine II and D50 W, with added sodium and potassium chloride. The mechanism permits the nurse to specify the composition of non-standard solutions quickly. Nurses need only select the proper additives from the list and enter the amounts.

FLUID ENTRY

1 ID# 111111111 TESTING
2 TIME 1100 18 JUN
3 WEIGHT(KG) 70.00 AT 0800

	TIME			
ENTRIES	0800	0900	1000	1100
4 D5/W	50	50	60	60
-> 1/2 NS + 30KCL	70	70	70	
6 FREAMINE 2 +ADD# 1	50	50	50	
7 WHOL BLOOD	125	-	-	
8 ISOCAL	40	50	50	
9 URINE	75	70	120	
10 NG DRN	120	-	-	
11 CHST 1-CLR	55	68	32	
12 RL	40	STOP	-	

99 STORE ENTRY

FOR INSTRUCTIONS PUSH '; GO'
ENTRY : PAGE 1 OF 1

Figure 3. Basic display used for nursing entry of intake and out-put volumes. A patient's current fluids are listed on the left, with three prior hours' entries displayed on the right. The arrow moves down the left margin as entries are made.

OPERATION OF THE SYSTEM

In order to have general applicability, the system was designed for nursing entry. In the 5600A system, entries are made
using hand-held 16-key or fixed, full ASCII keyboards. Displays
use video monitors or CRT terminals. An important design feature
is illustrated in Figure 3, the basic entry form. Each of the displays in the system is accompanied by a set of instructions on a
second display. Novice users may get instruction on how to proceed
at any stage; thus, the system is to some extent
self-instructional. A second capability designed into the system
to make it acceptable is that it permits experienced users to take
short-cuts to speed their tasks.

Another important design feature illustrated by Figure 3 is
that the volume entry form looks like presently used manual flow
sheets. When nurses make entries on manual flow sheets, they can
scan prior entries both to detect significant changes and to verify
past and present entries. The computer entry form uses this important clinical monitor. At the time that a new entry is made, three
prior hours' entries are retrieved from the patient's database and
displayed. The nurse may then enter the volume of each fluid administered or output, stop fluids no longer appropriate, or add new
fluids from the fluid list (Fig. 1).

When the nurse has entered the volume of each of a patient's
current intakes and outputs and indicates that the data are to be
stored, the system multiplies the volume of each fluid by its constituent concentrations. The totals are added to running sums
stored for each hour, nursing shift, day, and cumulative ICU admission. In order to keep the system's files from overflowing with
unnecessary data, the I/O system selectively purges outdated hourly
records at intervals definable by users.

After data have been stored, they may be printed or displayed
in a series of user-defined reports. Because the layout of the
files used for storing data depends on the definition of the constituent table which can be specified by users, report layouts similarly had to be configurable by users. We developed a special report generator for the I/O system which permits users to define reports formats using the regular Hewlett-Packard text editor, provides both horizontally and vertically formatted reports, and supports both printers and display devices. Reports can consist of
multiple pages, and can be paged forwards and backwards in time to
enable review of the entire admission. Examples of two pages are
shown in Figures 4 and 5. Figure 5 is interesting because it illustrates one of the potentially most important, but least well accepted, features of the system. The system can automatically calculate balances of sodium, potassium, and chloride. These balances
have not been routinely available to physicians in the past because

```
111111111 TESTING                    20 JUN 1155
```

FLUID VOLUME SUMMARY - JUN 18
                                0 SHIFTS MISSING

INTAKE(CC)	1400	2200	0600	24-HR	1-DAY
IV	740	1340	1550	3630	3630
WHOLE BLOOD	125	0	480	605	605
PACKED RBC	0	0	0	0	0
PLATELETS	0	0	0	0	0
PLASMA/ALB	0	0	0	0	0
ORAL	140	0	0	140	140
TOTAL INTAKE	1005	1340	2030	4375	4375
OUTPUT (CC)					
URINE	345	980	810	2135	2135
NG+VOMITING	240	423	483	1146	1146
CH+AB CLEAR	175	123	87	385	385
CH+AB BLOOD	0	0	0	0	0
BLOOD	0	0	0	0	0
DIARRHEA	0	0	0	0	0
OTHER	0	0	0	0	0
TOTAL OUTPUT	760	1526	1380	3666	3666
NET IN-OUT	245	-186	650	709	709

```
FOR INSTRUCTIONS PUSH ; GO        PAGE 1 OF 3
:
```

Figure 4.   The volume summary display.   Items   on
the report may be configured by users.

```
111111111 TESTING                    20 JUN 1201
```

ELECTROLYTE/NUTRITIONAL SUMMARY - JUN 19
                 0 SHIFTS MISSED

	24-HOUR			2-DAY			
	IN	OUT	NET	IN	OUT	NET	UNIT
NA	242	175	67	471	318	153	MEQ
K	76	112	-37	157	183	-26	MEQ
CL	290	299	-10	575	539	37	MEQ
NITR	8	4	4	14	5	9	GM
CA	1			7			MEQ
MG	1			4			MEQ
HCO3	0			0			MEQ
HPO4	12			23			MEQ
SO4	0			0			MEQ
BASE	55			91			MEQ
HGB	0			0			GM
CAL	1550			2969			CAL
PROT	47			92			GM
CARB	390			728			GM
FATS	3			12			GM

```
FOR INSTRUCTIONS PUSH ; GO       PAGE 3 OF 3
:
```

Figure 5.   The   electrolyte/nutritional   summary
display.   Items   on   this   report are selectable.   The
system automatically calculates both daily and   cumulative electrolyte and nitrogen balances.

they were too time consuming to calculate.   Thus,   physicians   are
faced with new data, the balances, which they have little basis for
interpreting.   Like water in a brackish reservoir with streams   and
tide   flowing   in   and   out,   where the net exchange determines the
level and salt concentration of the reservoir, we know that cumulative electrolyte balances coupled with water balance ultimately determine serum electrolyte levels.   When we have had sufficient   experience   in interpreting them, we feel certain that they will provide important information for care.

SUMMARY

The fluid balance system, begun over four years ago, typifies
the process of stepwise refinement in software development.   As new
user or medical requirements emerged, the system   evolved   to   meet
them.   It is still evolving.   Plans are being made to interface the
system to on-line fluid infusion pumps and drainage sensors so that
nurses, in some cases, will be relieved of the burden of making all
entries.   Further, we will be using the fluid entries as   data   for
automated decision making regarding fluid management and cardiopulmonary function using the AID system.(8) The fluid   balance   system
was   completed   about a year ago, was fully documented, and now has
been released and distributed generally for use by Hewlett-Packard.
The   system   is used routinely for care in some ICU's served by our
system and is a cornerstone of our clinical research program.

REFERENCES

1. WARNER, H.R.: Computer-based patient monitoring. In R. W. Stacy and B.D. Waxman (Eds.), Computers in Biomedical Research. Vol. 3. New York: Academic Press, 1969. Pp. 239-251.

2. REICHERTZ, P.L.: Medical School of Hannover Hospital Computer System (Hanover). In M.F. Collen (Ed.), Hospital Computer Systems. New York: Wiley, 1974. Pp. 598-661.

3. KLOEVEKORN, W.P., RICHTER, J., MENDLER, N. AND SEBENING, F.: Computer assisted intensive care monitoring. In Computers in Cardiology, IEEE Press, 1976. Pp. 237-241.

4. SIEGEL, J.H. AND STROM, B.L.: An automated consultation system to aid the physician in the care of the desperately sick patient. In R.W. Stacy and B. D. Waxman (Eds), Computers in Biomedical Research. New York: Academic Press, 1974. Pp.115-134.

5. SIEGEL, J.H., FICHTHORN, J., MONTEFERRANTE, J., MOODY, E., ET AL: Computer-based consultation in "care" of the critically ill patient. Surgery 80: 350-364, 1976.

6. MOORE, F.D.: Metabolic Care of the Surgical Patient. Philadelphia: Saunders, 1959.

7. BRIMM, J.E., O'HARA, M.R., AND PETERS, R.M.: An adaptable system for keeping intake/output records for fluid management in the critically ill. Second Annual Symposium on Computer Applications in Medical Care, Washington, D.C., November 1978. Long Beach, CA: IEEE Press, 1978. Pp. 620-624.

8. BRIMM, J.E., JANSON, C.M., PETERS, R.M. AND STERN, M.M.: Computerized ICU data management. Proc. First Annual International Symposium Computers in Critical Care and Pulmonary Medicine, Norwalk, Connecticut, May 24-26, 1979.

DETERMINATION OF PULMONARY HYDRATION BY MEANS OF TRANSTHORACIC IMPEDANCE MEASUREMENTS

G. Tempel, B. V. Hundelshausen

Institut fur Anaesthesiologie der Technischen
Universitat Munchen
Klinikum rechts der Isar
(Director: Prof. Dr. E. Kolb)

The admission to anaesthesiological intensive care units is most often required by respiratory difficulties and pulmonary malfunctions. Even today one of the main causes of death among these patients remains the intractable respiratory insufficiency. Undoubtedly – shock be it a septic or hypovolemic traumatic shock – is one of the major factors in the development and perpetuation of pulmonary complications. Nowhere in the body the manifestations of shock can so easily be observed as in the lungs, where clinical findings, x-ray visualization and physiological measurements allow the concomitant interpretation of functional and morphological data. Independent of the basic illness a uniform clinical, pathophysiological and morphological pattern can be seen in the course of shock complicated by the insufficiency of pulmonary gas exchange. Depending on the course of the illness several well defined phases can be identified and are easily mistaken for distinct pathological entities. Despite the great number of terms used for shock induced respiratory malfunction I would like to emphasize the uniformity of this clinico-pathological process. Moreover this lack of a common name for a common problem reflects our lack of understanding of the underlying etiologic and pathogenetic process. One fact we know for sure is that time itself plays a vital role in the development of the so called adult respiratory distress syndrome. This is shown by the well known clinical and pathological studies. The first figure originally published by Mittermayer and coworkers 1976 demonstrates, that better shock treatment during the last decades resulted both in a considerable gain in the time of

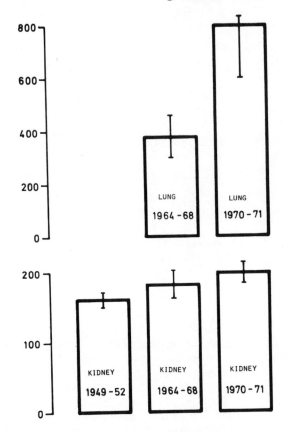

Fig. 1. Lung and kidney weight of patients dying in shock. The average length of survival increased from 4 to 11 days during the years 1949 - 1971. Ch. Mittermayer adn H. Joachim, Z. Rechtsmed. 78, 1 (1976).

survival and in gross morphological alterations unkown before. The lungs of patients having survived shock for 11 days had almost twice the weight of the ones who succumbed already after 4 days. It could be proven by morphometric methods, that the weight incre- ase is due to an interstitial edema and not to an increase of in- travascular volume - Joachim and coworkers 1976. These well docu- mented clinicopathological findings opened the road to additional therapeutic possibilities. Dyspnea, hypoxemia and interstitial edema the diagnostic features which are consistently found in shock induced respiratory dysfunction fit well together and link clinical findings to biochemical measurements, pathophysiological concepts, and morphological observations. Why the interstitial edema devel- ops, and how it does so, is not easily understood. It may well be, that the body itself has lost control over one of its homeostatic mechanisms, namely the physiological ability of the lung to store a fluid excess in the low pressure pulmonary circulation. The point where this adaptive mechanism turns from physiologic intravascular filling to interstitial hyperhydration is not easily marked. Clinical findings and x-ray examinations don't lend us a hand in this early phase. The main finding, the relative hypoxemia, can be viewed in different ways. I therefore would like to outline here the role of transthoracic impedance cardiography as a diagnostic tool in the assessment of impending adult respiratory distress and as an effective means in controlling therapeutic interventions.

The principle of impedance cardiography is based on the changes of bioelectrical resistance within the thorax in dependence on the fluid content. The impedance of thorax segment enclosed by two electrodes is dependent on the corresponding tissue. Ionized fluids have a comparatively low resistance, while air and tissue are highly resistant. The impedance of bones and fat tissue, lung and heart is constant. Changes of impedance in the thorax are caused by changes in the content of water and air. If the measurement is done at the end of an expiration, fluid shifts are responsible for changes in impedance.

Originally this method, first described by Kubicek and coworkers 1966, was used for the non-invasive measurement of cardiac output. With simultaneous recording of cardiac sounds and changes of transthoracic impedance data are available to calculate the stroke volume by a formula derived by Kubicek and coworkers 1970. The opinions of various investigators on the validity of the calculation of the stroke volume by this method differ widely. In our investigations the determination of stroke volume was of only secondary importance. Rather we have tried to detect changes in degree of hydration of the patient by measurement of mean thoracic impedance.

Two circumferential self-adhesive aluminium tape electrodes are fixed around the patient's neck and at the level of the thoracic inlet to measure the impedance of a high-frequency alternating current. Two other electrodes are positioned around the trunk at the level of the xiphoid and about 2cm below, as shown in a schematic diagram figure 2. An electrical field is established between the two outer electrodes, the two inner electrodes recording the drop of voltage in the defined segment of the thorax. The voltage difference is proportional to the transthoracic electrical impedance which is observed from a digital voltmeter. We use the Impedance Cardiograph Minnesota IFM 304 A, which transmits a high frequency (100 kHz), low current (4 mA) and which shows digitally the mean body impedance (Z_o). All measurements are carried out with the patient supine at the end of a normal expiration to obtain comparable results.

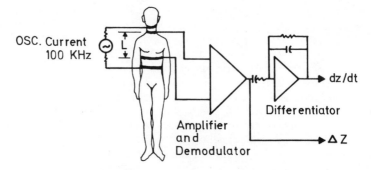

Fig. 2   Schematic diagram of impedance cardiography.

MEAN BODY IMPEDANCE

DATA FROM LITERATURE		
ADULTS	25	OHMS
OWN MEASUREMENTS		
CHILDREN UP TU 10 YEARS (N = 10)	26 ± 1.6	OHMS
ADULTS 20 - 40 YEARS (N = 23)	21.8 ± 1.8	OHMS
ADULTS MORE THAN 60 YEARS (N = 20)	24.2 ± 2.2	OHMS

Fig. 3.   Mean values of transthoracic impedance in healthy subjects.

First we determined the normal mean values for healthy persons, which have been found previously to be 25 ohms (Pomerantz, and coworkers 1970. In 20 children the mean impedance was approximately 26 ohms; in 30 men up to 40 years we found approximately 22 ohms and in 20 elderly men it was approximately 24 ohms. (Figure 3). The higher impedance in the children and elderly men can be explained by a higher content of air in relation to tissue. There was no correlation between the circumference of the thorax and the mean thoracic impedance.

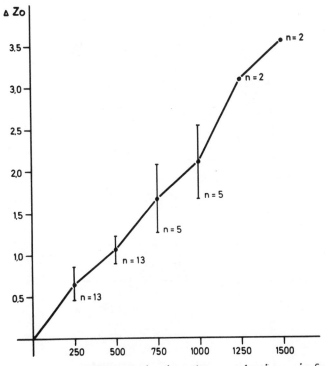

Fig. 4. Transthoracic impedance during infusion of isotonic electrolyte solutions; infusion of 500 ml resulted in a decline of impedance of about 1 ohm. n = Number of measurements.

The next figure illustrates changes of transthoracic impedance Z_o during infusion of 1500 ml of isotonic electrolyte solutions. There is a linear relationship between the amount of infused fluid and the decline of Z_o. Infusion of 500 ml of isotonic electrolyte solution is associated with a decline of impedance of about 1 ohm – figure 4. With a similar group of volunteer test subjects Z_o was recorded during a forced diuresis induced by furosemide. Here, too, a relationship between fluid shifts and Z_o was apparent – figure 5.

These results in healthy subjects indicated that the measurement of mean thoracic impedance may be useful for the diagnosis of fluid accumulation in patients with respiratory insufficiency.

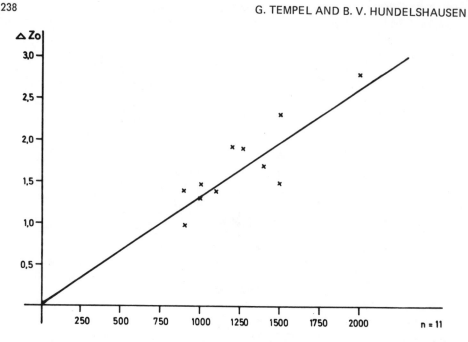

Fig. 5.   Changes of impedance during enforced diuresis with Furosemide;   diuresis of 1 liter changed
the mean thoracic impedance by approximately 1, 5   ohm.
n =11.

Figure 6 shows a record of a 77 year old patient with exogene-
ous hyperhydration after gastrectomy. After admission to the Intensive Care Unit, the patient's Zo was 18.8 ohms, the central venous pressure was 17 cm water and respiratory support was necessary.
A forced diuresis was started and parallel with a loss of 5 liters
of fluid the central venous pressure dropped and there was marked
increase in Zo. After normalization of hydration ventilatory support was no longer necessary and the patient could be weaned from
the respirator.

Now I would like to present the clinical course of a boy, aged
16 who was admitted, because of septic shock complicated by respiratory insufficiency, to out intensive care unit, Sept. 26th 1978
– figure 7.

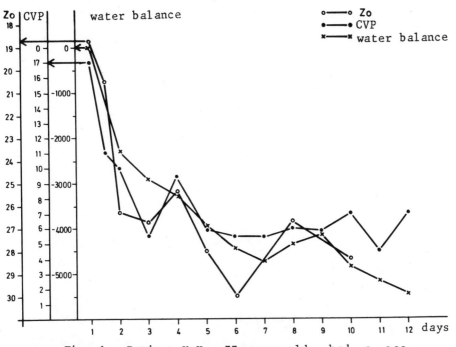

Fig. 6. Patient K.K., 77 years old, had a gastrectomy, hyperhydration and respiratory insufficiency.
On admission the central venous pressure (CVP) was 17
cm water and the mean impedance (Zo) was 18.8 ohms.
Enforced diuresis was started immediately and resulted
in a fluid loss of 4 liters in 5 days, a drop in central venous pressure and an increase of mean impedance
to 28 ohms. o, Zo; ●, CVP; X, water balance.

Nine days before he had suffered a craniocerebral concussion,
a fracture of the olecranon, ankle and the right hand and had developed a hemarthros of the knee joint. Originally awake, he gradually turned unconscious and showed signs of a marked respiratory
and circulatory insufficiency. On laparotomy a necrotizing acalcu-
lar cholecystitis was found. Treatment consisted of Dopamine infusions, controlled respiration and forced diuresis. Figure 7 shows
the water balance, the thoracic impedance Zo, the alveolar-arterial
oxygen tension difference AaDO2 and the serum creatinine during the

H.K., MALE PATIENT, 16 YEARS

17.9.78:   ROAD ACCIDENT
           HEAD INJURY
           FRACTURED OLECRANON
                   -OS METACARPALE
                   -ANKLE
           HAEMARTHROS

26.9.78:   SEPTIC SHOCK
           SEVERE RESPIRATORY DISTRESS

           ADMISSION TO THE ANAESTHESIA
           INTENSIVE CARE UNIT:
           INTUBATION, CONTROLLED RESPIRATION,
           FORCED DIURESIS, DOPAMINE INFUSIONS.

DATE	WATER BALANCE	Z.(OHMS)	$AaDO_2$(MMHG)	KREAT.(MG%)
26.9.78 14.00 H	---	19,1	294	2,6
26.9.78 18.00 H	-1500	20,8	150	2,5
27.9.78 10.00 H	-3400	23,3	46	2,4
28.9.78	-3300	23,2	42	1,5

Fig. 7. Clinical data, water balance, thoracic impedance, AaDO2 and serum creatinine during the first two days after admission to intensive care unit.

initial phase of treatment. The respiratory insufficiency was quickly reduced under the therapeutic regiment described above. Note the contemporaneous increase of the thoracic impedance Zo and the negative waterbalance, and the ensuing decrease of AaDO2.

We are confronted with quite similar problems during the treatment of multiple injured patients having undergone a trauma complicated by shock.

Figure 8 shows how AaDO2 changes with the waterbalance and transthoracic impedance during the treatment of 6 multiple injured patients. Thoracic hyperhydration was demonstrated by thoracic impedance measurements in all patients. All patients were admitted to our unit because of high grade respiratory insufficiency. There was marked improvement of respiratory function in all patients after the institution of controlled respiration and forced diuresis. All patients survived.

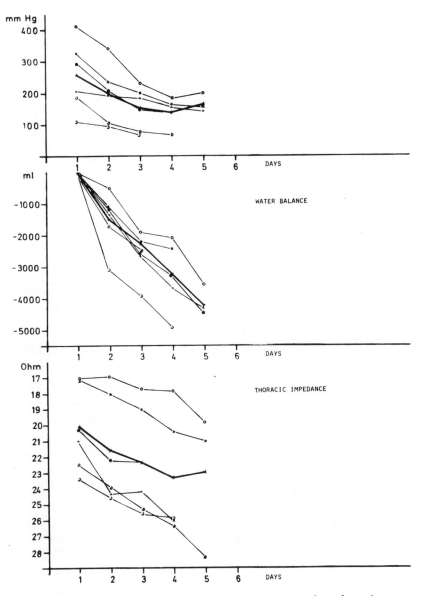

Fig. 8.   AaDO2, water balance, thoracic impedance
of 6 multiple injured patients during the posttraumatic
days 1 - 5.

If we try to base our therapeutic approach to postoperative
and posttraumatic respiratory insufficiency on sound facts, some
observations and statements have to be accepted.

The clinical course of respiratory insufficiency shows its characteristic phasic behavior. Hypoxemia is due to both an increase of intrapulmonary shunt and a decrease of alveolar oxygen diffusion. Interstitial edema is already seen in the early phase of respiratory distress. Some type of shock preceeding respiratory insufficiency can be found in each patient. Instant and sufficient shock treatment remains the most effective preventive measure. We have found mechanical ventilation and diuresis induced negative water balance to be of great value in the prevention and treatment of postoperative and posttraumatic respiratory insufficiency.

REFERENCES

1. JOACHIM, H., VOGEL, W., MITTERMAYER, CH.: Untersuchungen zum Phanomen der Schock lung. Rechtsmedizin 78, 13, (1976).

2. KUBICEK, W.G., KARNEGIS, I.N., PATTERSON, R. P., WITSOE, D.A., MATTSON, R.H.: Development and evaluation of an impedance cardiac output system. Aerospace Med. 37, 1208, (1966).

3. KUBICEK, W.G., PATTERSON, R. P., WITSOE, D.A.: Impedance cardiography as a nonivasive method of monitoring cardiac function and other parameters of the cardiovascular system. Ann. New York Academy: 170, 724 (1970).

4. MITTERMAYER, CH., JOACHIM, H.: Zur Pathomorphologie der Intensivbehandlung. Z. Rechtsmedizin 78, 1, (1976).

5. POMERANTZ, M., DELGADO, F., EISEMANN, B.: Clinical evaluation of transthoracic electrical impedance as a guide to intrathoracic fluid volumes. Ann. Surg. 171, 686, (1970).

ON-LINE MEASUREMENT OF RESPIRATORY GAS EXCHANGE: ROLE OF VO2 AS DETERMINANT OF O2 DELIVERY

R.W. Patterson and S.F. Sullivan

Department of Anesthesiology
University of California School of Medicine
Los Angeles, California 90024

We have put into practice a portable system for continuous non-invasive monitoring of oxygen consumption ($\dot{V}O2$) and of the respiratory quotient (R). These values have been useful in elucidating several abstruse causes of decreased arterial oxygenation as well as being useful in following trends and asymptotes of normal and abnormal respiratory physiology, and as guidance in the therapy of critically ill patients.

For example, in patients with chronic renal failure a consistent fall in arterial oxygen tension (PaO2) occurs shortly after initiating hemodialysis and is at times associated with serious cardiovascular symptoms. Some investigators [1], assuming no change in PAO2, implicate agglutinated white blood cell obstruction of pulmonary capillaries and consequent pathological interference with intrapulmonary gas exchange as causative of the (assumed) increased alveolar-arterial oxygen gradient, D(A-a)O2. Moving our equipment to the Dialysis Unit permitted continuous measurement over a six hour period of each patient's minute ventilation ($\dot{V}E$), mixed expired gas tensions (PEO2, PECO2), $\dot{V}O2$, $\dot{V}CO2$ and R (Figure 1). From these measured values we derived the alveolar ventilation, $\dot{V}A$, and the alveolar oxygen tension, PAO2. Our measured values for R decreased, PACO2 did not change, $\dot{V}E$ decreased. The decrease in R (from an average .85 to 0.5) and the decrease in $\dot{V}E$ (average 1.2 liter/min.) combine to reduce PAO2. The D(A-a) O2 did not change, and therefore PaO2 is reduced concomitant with PAO2 (figure 2).

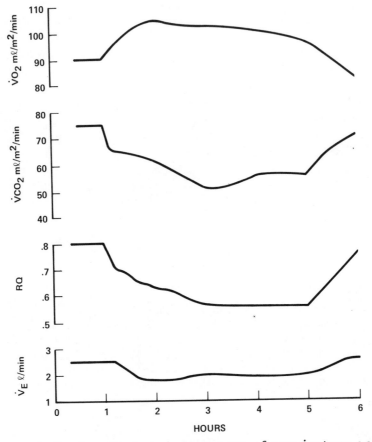

Figure 1.   Six hour measurement of respiratory gas exchange in a representative patient during hemodialysis, hours 1-5.

Thus, we demonstrated that the normal regulation of pulmonary gas exchange explained the hypoxemia measured in these patients without requiring the additional factor of intrinsic pulmonary dysfunction (2). This distinction provides the basis for proper therapy and prophylaxis.

A further illustration of our interests involves a consideration of the potential for disparity between oxygen utilization and oxygen availability during crucial periods in the operating room, recovery room, and ICU. Moving the equipment to the operating room has allowed us to document that change in the determinants of ventricular oxygen demand (heart rate, myocardial contractility, preload, and afterload) is reflected in $\dot{V}O_2$ (fig. 3). Though the period of anesthetic induction is rigorously assessed in patients

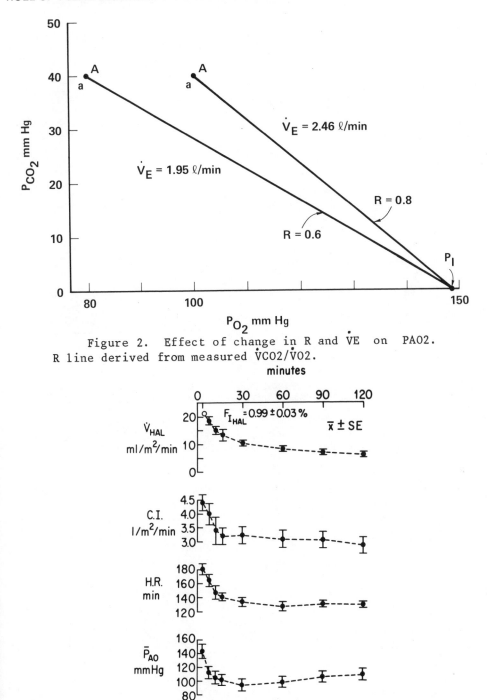

Figure 2.   Effect of change in R and $\dot{V}E$ on   PAO2.
R line derived from measured $\dot{V}CO2/\dot{V}O2$.

Figure 3.   The rate of decrease in heart rate, arterial blood pressure, and cardiac output parallels the
rate of anesthetic uptake. $\dot{V}O2$ decreases in a   similar
manner.

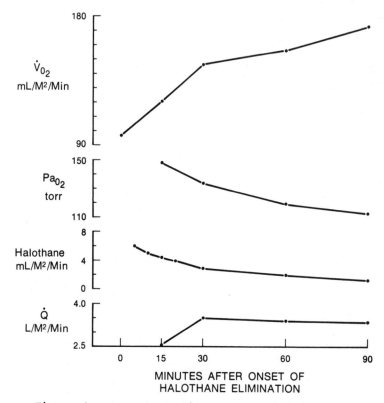

Figure 4.   As anesthetic agent is eliminated, in a
patient following open-heart surgery, there is a marked
increase in V̇O2 associated with a decrease in PaO2.

with coronary artery disease to assure a quantity of oxygen (arterial   oxygen   content,   CaO2,   X myocardial blood supply) adequate to
the oxygen requirement of the ventricular   muscle,   emergence   from
anesthesia can be equally hazardous.   In the recovery room inadequate analgesia, for example, may be associated with an increase in
V̇O2 sufficient to depress arterial oxygenation and thus tissue oxygen supply (fig. 4).

Continuous measurement of respiratory gas exchange is required
to   clearly delineate basal states from activity, steady-state conditions from transients, and changes in body gas stores   from   true
metabolic   changes (3,4);   all of which have been major sources of
error with discrete discontinuous measurement.   Steady state gas exchange conditions are defined as a constant ratio of CO_2 production
to oxygen consumption (R), or alternatively, a   constant   ratio   of
inert   gas   exchange (ratio   of   expired   to   inspired   nitrogen,
FEN2/FIN2);   both indices are continously recorded on   the   bedside
display terminal.

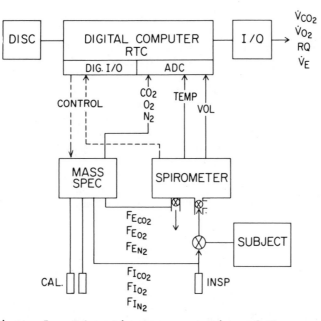

Figure 5. Schematic representation of the system employed to obtain respiratory gas exchange measurements on spontaneously breathing or mechanically ventilated subjects.

In addition to the capability of differentiation of changes in body gas stores from changes in metabolic states, long-term real-time determination of $\dot{V}O2$ in the mechanically ventilated or spontaneously breathing intubated patient during a change in environmental conditions or during subject movement dictates a non-invasive method. We have chosen to determine $\dot{V}O2$ from accurate measurement of expired gas volume ($\dot{V}E$) and mixed expired gas concentration using an open circuit approach. The essential relationships in gas exchange are represented by the following three equations:

$$\dot{V}CO_2 = \dot{V}E \left[FECO_2 - \left(FICO_2 \cdot \frac{FEN_2}{FIN_2} \right) \right] \qquad (1)$$

$$\dot{V}O_2 = \dot{V}E \left[\left(FIO_2 \cdot \frac{FEN_2}{FIN_2} \right) - FEO_2 \right] \qquad (2)$$

$$RQ = \frac{\dot{V}CO_2}{\dot{V}O_2} \qquad (3)$$

The basic elements of the system employed to obtain these data are shown in figure 5. The subject, who may be spontaneously breathing or mechanically ventilated, is connected to a nonrebreathing valve system which separates inspiratory gas from expiratory gas. Exhaled gas is directed into an electronic volume displacement spirometer (Model 220, Spirometer, Cardiopulmonary Instruments, Inc. Houston, Texas) whose 'full' volume is operator adjustable and typically is set to collect approximately one minute volume of expired air each cycle. Logic integral to the spirometer performs two function: 1. signals the time of spirometer filling, during which period the mass spectrometer is utilized in calibration and analysis of the inspired gas concentrations; 2. determines end expiration so that as the pre-set collection volume is approached, gas continues to be collected until end expiration, i.e., there is no cut-off during midbreath. At this time the spirometer volume is logged and the valves opened permitting the subject to exhale freely to atmosphere while the spirometer volume is dumped. During the dump, the mass spectrometer (Model 1100, Mass Spectrometer, Perkin-Elmer Corp., Pomona, Calif.) whose multiple inlet ports are under computer control according to the spirometer phase, analyzes the mixed expired gas concentrations. The inspiratory gas, which may be room air or gas obtained from a premixed compressed gas cylinder, is analyzed as the spirometer is being filled. Calibration gas mixtures, from precisely analyzed cylinders directly referable to gravimetrically prepared standards (Liquid Carbonics Specialty Gas Laboratory, Los Angeles, California) which constitute the reference base for all gas analysis in our laboratory, are similarly sampled on each spirometer filling cycle providing a two-point calibration for each gas of interest.

Calibration, volume measurement, respired gas analysis, time-keeping and associated functions are performed under control of a comprehensive computer program implemented in BASIC and run on a laboratory minicomputer system (NOVA 3, Data General Corp. Southboro, Mass.) Upon completion of the computations utilizing the relationships in Equations 1,2,3, the bedside display terminal provides the clinician with constantly updated information on the subject's ($\dot{V}E$ (BTPS), $\dot{V}O_2$ (STPD), $\dot{V}CO_2$ (STPD), R and FEN_2, as well as other directly obtained and derived cardiopulmonary data.

The reliability of the system over a 2-hour period has been measured using a model with constant ventilation and constant inspired and expired gas concentration. Individual data points for $\dot{V}E$, FIO_2, FEO_2, $FECO_2$, and $\dot{V}O_2$ were subjected to statistical analysis. Individual variation in gas concentration measurement is minimal. For example, FIO_2 (%) measured at 20.94 +- 0.0113 ($\overline{X}$+-S.E.), FEO_2 (%) at 15.75 +- 0.0064, $\dot{V}E$ (L/min) at 7.16 +- 0.0023, and $\dot{V}O_2$ (mL/Min) at 332.9 +- 0.81. These analyses indicate a coefficient of variation for $\dot{V}O_2$ of approximately 2.9%.

The use of a volume displacement gas collection device eliminates phase lag and sensor response time problems associated with
integrated flow methods. Although the spirometer imposes more
equipment bulk at the bedside, the greater simplicity, accuracy and
long term reliability afforded by this method justifies the
trade-off in obtaining steady state measurements.

It is expedient to conclude this brief presentation with a
description of the usefulness of $\dot{V}O2$ measurement in evaluating a
common therapeutic dilemma: - the uncertainty in assessing the direction and magnitude of effect when numerous factors and treatments are combined at one time. The total amount of oxygen available each minute ($\dot{Q}T$. CaO2) is equal to the quantity of oxygen
delivered form the lung each minute (the product of the pulmonary
blood flow effectively participating in gas exchange and the pulmonary capillary oxygen content, (QT - QS . Cc'O2) combined with the
oxygen contribution of mixed venous blood passing from the right to
the left side of the circulation without gas exchange (QS . $\overline{Cv}O2$).
In the Fick equation, mixed venous oxygen content ($\overline{Cv}O2$) is equivalent to CaO2 - ($\dot{V}O2/\dot{Q}T$). Substituting this term into the first
statement provides the following relationship:

$$CaO_2 = Cc'O_2 - \left[(Q_S/Q_T) \cdot \frac{\dot{V}_{O2}}{Q_T - Q_S} \right] \qquad (5)$$

i.e., CaO2, the arterial oxygen content, representing oxygen delivery, is dependent upon a term heavily weighted by ventilation parameters, intrapulmonary right to left shunting of blood, total body
oxygen consumption, and cardiac output. From this relationship the
pivotal role of $\dot{V}O2$ becomes evident. The terms on the right hand
side are subject to a myriad of changes in critically ill patients.
The closer cardiac, circulatory, and pulmonary functions approach
the limits of their compensatory reserve the more CaO2 is directly
dependent upon $\dot{V}O2$.

Basic to the present discussion is the fact that respiratory
alkalosis, and metabolic alkalosis as well (5), leads to an increase in $\dot{V}O2$. Thus, each of the right hand terms in th model of oxygen transport have previously been shown to be individually altered
by hyperventilation; an increase in Cc'O2 directly resulting from
the increased PAO2' and an increase in magnitude of calculated QS'
and increase in cardiac output and an increase in total body oxygen
consumption due to the decrease in arterial [H+]. The problem now
becomes: - What is the overall effect on arterial oxygenation when
a patient with atelectasis is passively hyperventilated? The ap-
portionment of the relative contribution of each determinant (Equation 5) to the final effect on CaO2 requires that all of the terms
be simultaneously evaluated. In an animal model of these conditions the institution of passive hyperventilation reduced right and

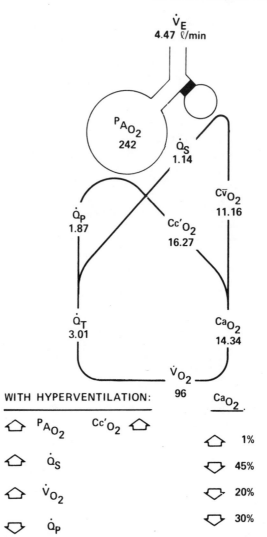

Figure 6. Representative study. Conditions;
bronchial blockade followed by passive hyperventilation. Increased measured oxygen consumption accounts
for 20% of the total decrease in arterial oxygen carrying capacity, CaO2.

left ventricular afterload, cardiac output increased and the resultant increase in V̇O2 accounted for 20% of the total decrease in arterial oxygen content. However, 45% of the total decrease in CaO2
was due to alkalotic attenuation of hypoxemic vasoconstrictor re-
flex, reflected as an increase in QS/QT (figure 6). Thus, the observed results differed from those anticipated. Accurate V̇O2 measurement contributes to this type of evaluation which allows confi-
dent recommendation for specific therapy.

REFERENCES

1 CRADDOCK, PR, FEHR, J, BRIGHAM, KL, KRONENBERG, RS AND JACOB, HS:
   Leukocyte-mediated pulmonary dysfunction in hemodialysis.
   N. Engl. J. Med. 296: 769-774, 1977.

2 PATTERSON, RW, NISSENSON, AR, MILLER, J, SMITH, RT, NARINS, RG,
   AND SULLIVAN SF: Hypoxemia and pulmonary gas exchange
   during hemodialysis. Submitted to J.   Appl.   Physiol.,
   1979.

3 SULLIVAN, SF, PATTERSON, RW, AND PAPPER, EM: Arterial CO2 tension   recovery   rates   following   hyperventilation.   J.
   Appl. Physiol. 21: 247-250, 1966.

4 SULLIVAN, SF, AND PATTERSON, RW: Variability in oxygen uptake
   during spontaneous ventilation in resting man. Fed.
   Proc. 28: 654, 1969.

5 PATTERSON, RW, AND SULLIVAN SF: Determinants of oxygen uptake
   during sodium bicarbonate infusion. J. Appl. Physiol.:
   Respirat. Environ. Exercise Physiol 45 (3):399-402,
   1978.

THE MEASUREMENT OF THE PULMONARY DIFFUSING CAPACITY BY A REBREATHING TECHNIQUE DURING CONTROLLED VENTILATION.

H. Burchardi, T. Stokke,
I. Hensel, H. Angerer,
W. Roehrborn

Institut fuer klinische Anaesthesie
der Universitaet Goettingen
Goettingen, West Germany

This study was supported in part by a grant from the Deutsche Forschungsgemeinschaft, Bonn.

Until now, measurements of the pulmonary diffusing capacity in intensive care medicine were not usual. They were unpopular because in presence of severe respiratory insufficiencies the results obtained by the routinely used methods often may be superimposed by various non-specific effects. Namely, in the presence of uneven ventilation the "single breath" and "steady state" methods will lead to an underestimation of the diffusing capacity.

KRUHOFFER and LEWIS et al. assumed that the rebreathing technique would offer more correct results in ventilation-perfusion disturbances than the other methods. To a certain degree this could be proved later by SCHEID et al. and by SOLVSTEEN.

The mathematical analysis of the rebreathing method has been substantially improved recently by ADARO et al. In order to use this method in intensive care medicine we modified the technique for measurements during artificial ventilation.

Here we will describe our method and present some preliminary results from animal experiments and clinical application performed by this technique.

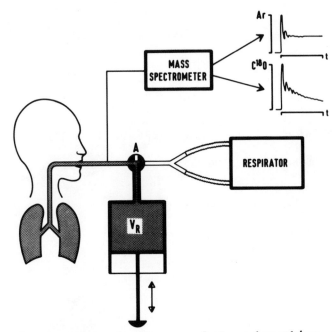

Fig. 1 Schematic diagram of the rebreathing procedure for measurements of the pulmonary diffusing capacity (DLCO) and the functional residual capacity
(FRC) during artificial ventilation. Rebreathing reservoir (VR) is a 1-liter-supersyringe connected to the
patient's lung by a 3-way-stopcock (A). Gas concentrations (Ar and C180) are measured continously by mass
spectrometer and recorded.

Method:

    A 1-liter-supersyringe as rebreathing reservoir is filled with
a test gas mixture containing 0.2% C180 (a stable isotope of CO)
and 10% Argon (fig.1). The lung system of the artificially ventilated patient will be connected to the syringe at the end of an expiration. By rapid manual rebreathing to and fro, the equilibration of the gases in the whole system is performed in about 10 -20
seconds. The test gas concentrations are measured continuously by
a mass spectrometer (PERKIN -ELMER MGA 2000).

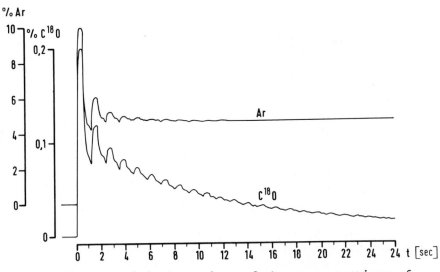

Fig. 2   Original tracings of the concentrations of
Argon (Ar) and C18O during the rebreathing maneuver.

Before each measurement, a separate rebreathing maneuver with
the expired gas mixture will be performed in order to determine the
"back pressure" of C18O and Ar in the body.   The total maneuver is
performed in less than 3 minutes without any severe disturbance,
even in critically ill patients.

Analysing process (see appendix):

After a period of gas mixing, the equilibrium is achieved
(Fig. 2).   Based on the Argon concentration during this equilibration period the functional residual capacity (FRC) can simply be
calculated.

The determination of the diffusing capacity (DLCO) is more
complex: During the period of equilibration and before a recirculation of CO, the slope of the semilogarithmic concentration curve
represents the rate constant for CO uptake (fig.3.) Curve evaluation and mathematical procedure can be substantially facilitated by
semi-automatical analysis with a HEWLETT PACKARD digitizer and calculator (A 9852).   At present we are studying procedures for complete automatic computation.

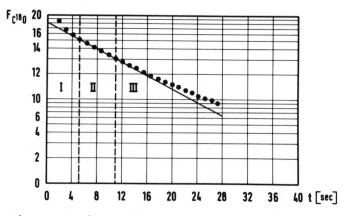

Fig. 3 Semilogarithmic plot of the C180 concen-
tration course during the rebreathing maneuver.
I = mixing period. II = equilibration period.
III = recirculation period.

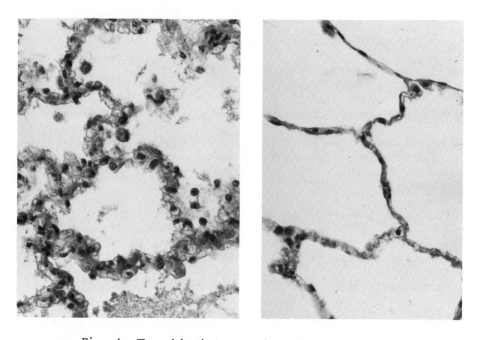

Fig. 4 Thrombin-induced disseminated intravascu-
lar coagulation in mini-pigs as a model for shock lung:
Right: marked interstitial edema after thrombin.
Left: normal lung. (GIEMSA, 560 X)

The results obtained by our method are accurate and reproducible: In animal experiments the variation coefficients for multiple measurements are $\pm$ 3.6% for DLCO and $\pm$ 4.4% for FRC respectively. During a steady state of several hours, the variation coefficients for measurements in each animal are $\pm$ 10.5% for DLCO and $\pm$ 7.6 for FRC respectively.(9)

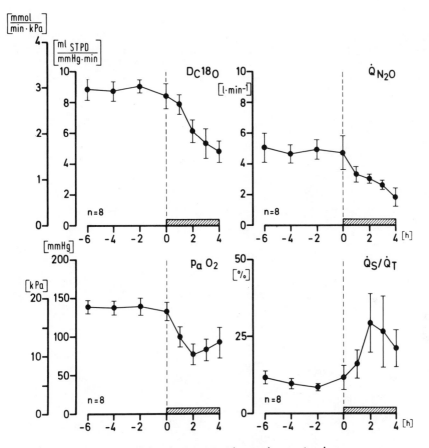

Fig. 5 Thrombin-induced disseminated intravascular coagulation in 8 mini-pigs as a model for shock lung: Upper left: Pulmonary diffusing capacity (DC18O) determined by rebreathing. Upper right: Pulmonary capillary perfusion QN2O. determined by rebreathing with N2O. Lower left: Venous admixture (QS/QT), calculated with the shunt formula. Mean values and standard deviations of the mean. Time scale (hours): 0 = Onset of the thrombin-infusion (150 U/kg/h).

Preliminary results:

Animal experiments

Disseminated intravascular coagulation induced by infusion of thrombin proved to produce lung changes very similar to the initial states of a shock lung (OLSSON et al.). Typical morphological changes are for instance a marked interstitial edema as seen in the beginning of a shock lung (Fig. 4). We induced disseminated intravascular coagulation in 8 mini-pigs by thrombin (150 U/kg/h) in order to study early changes of lung function in this shock lung model. The anesthetized mini-pigs received 40% oxygen by controlled ventilation during the whole experiment. After the onset of the thrombin infusion, a rapid and progressive deterioration of

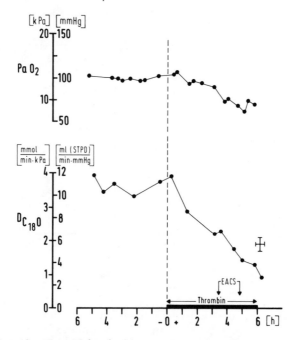

Fig. 6 Thrombin-induced disseminated intravascular clotting in mini-pig. Deterioration of diffusion by interstitial edema: Marked rapid and progressive decrease of the diffusing capacity (DC180) immediatedly after onset of thrombin, whereas the arterial pO2 reacts more slightly and with delay.

the pulmonary gas exchange function was obvious with a decrease of
arterial pO2 (Fig. 5). This progressive disturbance of oxygen exchange is caused on the one hand by an increase of venous admixture
(calculated with the shunt formula), on the other hand by a considerable decrease of the pulmonary diffusing capacity.

This effect of the diffusing capacity starts very promptly
after the onset of the thrombin. In two of our animals the decrease of the diffusing capacity began significantly earlier and more
pronounced than the changes in arterial pO2, which only decreased
more slightly and with delay (Fig. 6).

Thus we are of the opinion that some incipient deteriorations
of pulmonary gas exchange, like those for instance, which are typical for the initial period of shock lung under some conditions may
be detected earlier by determination of the diffusing capacity than
by measurements of the arterial pO2.

Further clinical obervations are necessary to prove evidence
this hypothesis.

Clinical investigations:

The following clinical case history may also support this assumption:

A 15 year old girl was admitted to our hospital after suicidal
poisoning with a high dose of paraquat (Fig. 7). Paraquat (GRAM-
MOXONE) is a very toxic herbicide causing typical progressive lung
changes (the so called "paraquat lung") with pulmonary edema and
later on a highly lethal interstitial fibrosis (COOKE et al.).   In
spite of intensive care medicine including gut lavage, peritoneal
dialysis and hemofiltration, the girl died 30 days later.   During
this period, the deterioration of the gas exchange became the main
problem: 4 days after her admission the alveolar-arterial oxygen
tension gradient (AaDO2) suddenly increased by a marked pulmonary
interstitial edema. Gas exchange improved temporarily under controlled ventilation with PEEP. But during the following days gas
exchange progressively worsened again with a rapid and marked decrease of pulmonary diffusing capacity.

The correlation between the DLCO and AaDO2 demonstrates the
fact that at lower values of AaDO2 (under 250 mm Hg) the decrease
of DLCO is very pronounced (Fig. 8).

This supports our hypothesis that diffusing capacity may be a
very sensitive parameter for early gas exchange deteriorations.

The extremely low values of the DLCO during the last days are
clearly explained by the postmortem histologic findings (Fig. 9)

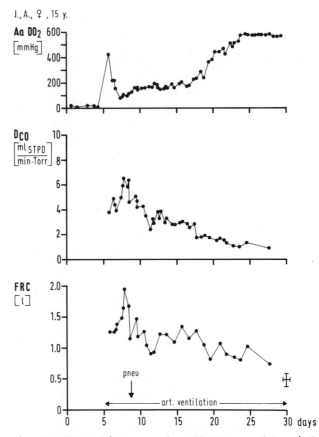

Fig. 7 Case history of a 15 year old girl with paraquat poisoning (see text). Controlled ventilation with PEEP after the 5th day; pneumothorax (pneu) at the 8 day, so reduction of PEEP became necessary. Died after 30 days. Changes in alveolar-arterial oxygen tension difference (AaDO2), in pulmonary diffusing capacity (DCO) and in functional residual capacity (FRC).

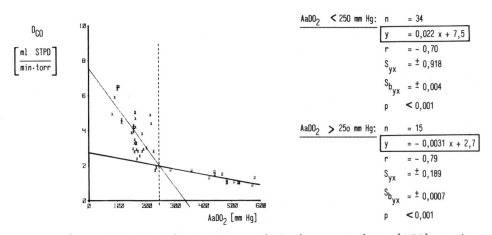

Fig. 8 Correlation between diffusing capacity (DCO) and alveolar arterial oxygen tension difference (AaDO2) from the case (Fig. 7): In AaDO2 values below 250 mm Hg deterioration of diffusion is very pronounced, whereas at AaDO2 over 250 mm Hg the gas exchange function may mainly be disturbed by other effects (like shunting).

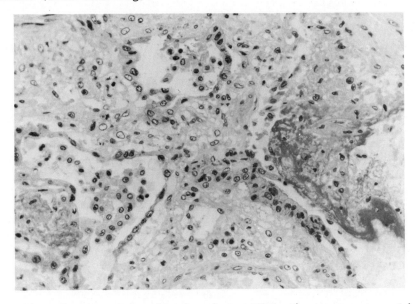

Fig. 9 Paraquat poisoning (case Fig. 7), postmortem histology of the lung: Severe interstitial fibrosis with proliferation of the fibroblasts and the pneumocytes. Hyalin membranes. (Trichrom GOLDNER, semi thin section) (Pathol. Institut, University of Goettingen, Prof. Kunze)

Discussion:

It has to be pointed out that interpretation of pulmonary diffusing capacity changes is still very difficult. It is a parameter affected by very complex influences especially under pathologic conditions: Regional inequalities in ventilation, perfusion and diffusion certainly may affect the results in some way (i.e. always with falsely too low values); but the results are less affected than with other techniques. SOLVSTEEN proved that the results are correct if DLCO is distributed proportionally to the lung volume; but DLCO will be too low if it is distributed proportionally to the alveolar ventilation.

An essential factor of pulmonary diffusing capacity is the pulmonary blood volume. So variations in pulmonary blood volume will result in changing diffusing capacity. The results will be further affected by changes in pulmonary capillary pO2 and by changes in hemoglobin concentrations. The reduction of 1 g% hemoglobin causes 6.3% decrease of DLCO (RIEPL).

Inspite of these problems, I feel that pulmonary diffusing capacity could offer supplementary information especially in early respiratory insufficiency - even if the interpretation may be complex.

We feel encouraged to perform further investigations.

Summary:

In 1976, ADARO et al. introduced an improved mathematical analysis for the estimation of the pulmonary diffusing capacity (DLCO) by means of inert gas with a rebreathing technique. We modified this technique for the application of measurements under controlled ventilation. The passive rebreathing maneuver is performed with a gasladen 1-1-supersyringe. The test gas contains 0.2% C18O and 10% Ar. The equilibration course of the gases is measured by mass spectrometry. The calculating process is done semi-automatically by electronic calculators. In case of repeated measurements it is necessary to estimate the raised back pressure of the inert gas in the body. This is done by a separate rebreathing maneuver prior to the measurement. The functional residual capacity (FRC) is also calculated by this method. The measurements are reproducible: Multiple measurements showed a coefficient of variation of $\pm$ 3.6% for DLCO and $\pm$ 4.4% for FRC resp. In 8 mini-pigs disseminated intravascular coagulation was induced by thrombin as a model for shock lung studies. A short time after onset of thrombin a marked and progressive deterioration of pulmonary gas exchange developed caused by a pulmonary interstitial edema. DLCO was very pronouncely decreased; in some animals it seemed to be a more sensitive parameter for beginning diffusion disturbances than pO2. This was also evident in a clinical case with paraquat poisoning leading to a progressive deterioration of diffusion by pulmonary edema and interstitial fibrosis.

Appendix

Calculation of the pulmonary diffusing capacity (D_{LCO}) determined by the rebreathing technique (ADARO and co-workers. 1976):

$$D_{CO_{RB}} = V_A \cdot B_g \cdot k_{CO} \cdot f \quad ml_{STPD} \cdot min^{-1} \cdot torr^{-1}$$

here is:

V_A = mean effective alveolar volume during rebreathing (determined by the equilibration of argon)

B_g = solubility of CO in the gas phase (o.523 mmol . min^{-1} . $torr^{-1}$)

k_{CO} = CO uptake (rate constant) during the equilibration period

f = correction factor for compensation of inhomogenous distribution in the whole rebreathing system:

$$f = 1 + \frac{V_R / V_A}{1 - k_{CO} \cdot V_R/\dot{V}_{eff}}$$

here is:

V_R = mean effective volume of the rebreathing syringe

$\dot{V}_{eff}$ = effective ventilation between rebreathing syringe and lung, calculated from the argon mixing process.

REFERENCES

1. ADARO, F., SCHEID, P., TEICHMANN, J. and J. PIIPER: A rebreathing method for estimating pulmonary DO2: theory and measurements in dog lungs. Respir. Physiol. 18, 43 (1973)

2. ADARO, F., MEYER, M, R.S. SIKAND: Rebreathing and single breath pulmonary CO diffusing capacity in man at rest and exercise studied by C18O isotope. Bull. europ. Physiopath. resp. 12, 747 (1976)

3. COOKE, N.J., D.C. FLENLEY, H. MATTHEW: Paraquat poisoning. Serial studies of lung function. Quart. J. Med. 168, 683 (1973)

4. KRUHOFFER, P.: Studies on the lung diffusion coefficient for carbon monoxide in normal human subjects by means of C(14)O. Acta physiol. scand. 32, 106 (1954)

5. LEWIS, B.M., E.J. HAYFORD-WELSING, A FURUSHO, L. C. REED: Effect of uneven ventilation on pulmonary diffusing capacity. J. Appl. Physiol. 16, 679 (1961)

6. LEWIS, B.M., T.H. LIN, F.E. NOE, E.J. HAYFORD-WELSING: The measurement of pulmonary diffusing capacity for carbon monoxide by a rebreathing method. J. Clin. Invest. 38, 2073 (1959)

7. OLSSON, P., K. RADEGRAN, J. SWEDENBORG; The platelet release reaction as a cause for functional pulmonary changes in disseminated intravascular coagulation. In: HABERLAND, G.L., D.H. LEWIS (Ed.): New aspects of trasylol therapy. Vol. 6. The lung in shock. p. 37. Schattauer, Stuttgart - New York 1973

8. RIEPL, G: Effects of abnormal hemoglobin concentration in human blood on membrane diffusing capacity of the lung and on pulmonary capillary blood volume. Respiration 36, 10 (1978)

9. SACHS, L.: Angewandte Statistik. Springer, Berlin - Heidelberg - New York, 1974

10. SCHEID, P., J. PIIPER: Diffusing in alveolar gas exchange. In: MUSHIN, W.W., J.W. SEVERINGHAUS, M. TIENGO, S. GORINI (ed.): Physiological basis of anaesthesiology. Piccin medical books, Padua, 1975, pp 71-94

11. SOLVSTEEN, P.: Lung diffusing capacity with particular reference to its determination in patients with uneven ventilation. Danish Med. Bull. 14, 142 (1967)

CLINICAL EXPERIENCE WITH A MICROCOMPUTER-BASED INTENSIVE CARE DATA MANAGEMENT SYSTEM

J.S. Augenstein, M.D., Ph.D., J.M. Civetta, M.D.,
G.F. Andrews, I. Lustig

Department of Surgery
University of Miami School of Medicine
Miami, Florida 33101

The creation of appropriate man-machine interfaces is critical if computers are to aid clinicians through high speed, highly flexible data organization. In theory, the computer, through data management approaches, could display the critical data for one or more patients on a single CRT. The same management scheme that provides instantaneous review for immediate therapy could provide answers to long range research questions. Data would not be duplicated nor would it exist on multiple forms. Communication between clinician and computer must be developed to provide smooth and reliable exchange of information. The overall approach in this work is to analyze the requirements of clinicians in clinical data management. Creation of computer systems and training in their use should provide enhancement of care delivery.

The intensive care unit (ICU) generates more data per patient unit time with more significant implications for patient survival than any other ward in the hospital. Computers have been used for more than 15 years in the ICU with the greatest success in monitoring data such as heart rhythm or respiratory status. Little work has been done on computerized management of the total data base. Most ICU computer systems duplicate activity by requiring charting on traditional records and making computer entries. The clinician is often required to deal with records distributed between handwritten and computer media with less integration than the handwritten records alone offer. Furthermore, limited CRT display capabilities often require multiple frames to display what a single handwritten record could have provided. Thus limited CRT affords less insight into the clinical situation than the traditional approach.

Computers have offered little to the intensive care clinician in complex data organization. Yet there are great demands for organized data not only in on-line decision making but also in meeting medico-legal, financial and research needs.

The work includes the following aspects: (1) developing a low cost, flexible computer system capable of total data base management in a surgical ICU; (2) developing techniques and using them to analyze the performance of the system in the clinical environment; and (3) evolving the computer system based on the data from the performance studies.

A distributed intelligence approach has been used to create a microcomputer based surgical intensive care data management system. The intelligence in the system is distributed between a microcomputer dedicated to each bedside and a central microcomputer shared amongst the bedside units. Both computers are Altair 8800A's which use floppy disks as the extended storage medium. The software is written in Level 5.0 BASIC with assembler language sub-routines. The bedside computer acquires data from the Spacelabs Alpha 9 cardiovascular monitor, drives a CRT at the head of the bed and accepts control and data inputs from a R-F linked handheld keyboard. This computer provides stand alone operations with: (1) multiparameter graded alarms, (2) cardiac and pulmonary calculations and (3) tabular and graphical review via its CRT. The central computer contains the extended management for fluid input-output data, medications, laboratory data and notes. Program development capabilities and hardcopy facilities also are included.

Much attention has been directed towards the integration and utilization of the system in the clinical environment. The system was designed to provide total patient data management. Examination of many existing computer systems revealed the dispersion of information between handwritten and computer generated records, much duplication of data entry, and little long term data management. Each of these was considered significant weakness but by necessity, was contained in early stages of the system. Human factors analysis techniques including time motion studies, social psychology techniques, questionaires and interviews were utilized. Via these tools the system's interface with nurses, physicians and technicians was examined. The time requirements and attitudes toward each of the nursing and physician functions with respect to data acquisition and review were first analyzed in the existing manual patient care system. Next, these same tasks were evaluated in the computerized system at various stages of development. A number of observations were clear: (1) The computer was poorly accepted when it required duplication of effort, particurary in data entry. (2) The time of data entry was reduced more than 50% with the use of a movable keyboard and head of the bed CRT. (3) A CRT at the head of the bed was not useful for most complex reviews. The patient support equipment blocked close scrutiny of the screen and most people would not read complex displays even on a 19" screen from more than a few feet. A CRT built into a desk at the foot of the bed functioned better for this purpose. (4) The system required directed branching tree structures for new and insecure users. For experts, the system allowed direct access and flexible displays capable of integrating many types of data. Most physicians preferred to use their own data combinations on the CRT rather than use the standard review displays. (5) Many physicians also preferred to use the traditional twenty-four hour handwritten record. Multiple CRT frames were required to present all relevant data included in that single sheet. However, handwritten records over twenty-four hours old were often lost or dispersed within a chart. The CRT display provided the most complete and flexible display of this data. It was felt that the computer system can reduce nursing clerical activity significantly.

In summary, a distributed intelligence microcomputer system was developed to provide total data management in a surgical intensive care unit. Using simple human factors and social psychology techniques, the utilization of the system was examined. This provided directions for development with greater system efficiency and user acceptance over time. This approach of flexible software and hardware plus scientific analysis of the system's use seemed to be critical in developing meaningful computer applications in the intensive care unit. It is hoped that this work can provide a guideline for the development of general utility clinical information systems.

CLINICAL EVALUATION OF AUTOMATED ARRHYTHMIA MONITORING AND ITS

INTEGRATION WITH RESPIRATORY MONITORING IN A GENERAL COMPUTERIZED

INFORMATION SYSTEM

Svante Baehrendtz, Johan Hulting,
Georg Matell, Mats-Erik Nygårds

Department of Medicine
Karolinska Institutet at Södersjukhuset
Stockholm

Department of Medical Informatics
Linköping University

At Södersjukhuset, Stockholm a medical intensive care unit (ICU) was started in 1955 by the late Gösta von Reis. This unit took an active part in the development of the "Scandinavian method" for the treatment of barbiturate poisoning. Later interest was also focused on myasthenia gravis. In the sixties the unit was doubled in size to incorporate facilities for coronary care. During these years the medico-technical unit began collaborating with the Departments of Environmental Medicine, Karolinska Institute and Medical Informatics, Linköping University. This work dealt with the development of new ventilators and also resulted in a computer program for the analysis of cardiac catheterization data (Nygårds et al 1976). In several pilot projects we studied different uses of computer technique in the ICU were studied. One was the successful use of an analog computer for blood pressure regulation in barbiturate poisoning with shock (Bevegård et al 1969). Following these pilot studies we decided to concentrate on arrhythmia surveillance since we regarded automatic non-invasive cardiac monitoring to be fundamental to successful intensive care.

An arrhytmia monitoring system is described below. This is followed by a description of how this system will be coordinated with an information system, including respiratory monitoring.

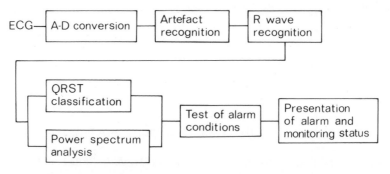

Fig. 1 Program organization in the arrhythmia detector. (Reproduced from Acta Med Scand, Suppl 630, with permission)

I. ECG Monitoring System

Our work on ECG monitoring was commenced in 1970 and resulted in new algo- rithms for ECG signal analysis. An 8-patient system was taken into clinical use in 1976 (Nygårds and Hulting 1979).

The program organization of the arrhythmia detector is show in Figure 1. After A/D conversion, artifact detection is performed permitting automatic rejection of low-quality ECG portions. An R wave is recognized by measuring the voltage differences between sample points of the ECG. A waveform recognized as a possible R wave is further analysed in a number of steps, using a feature extraction technique. Each QRST complex is mathematically represented from four orthonormal basis functions and a width parameter.

The waveform of each beat is classified with the aid of two discriminant functions operating on the waveform features and rhythm data. Similar beats are clustered. The classification of ventricular fibrillation (VF), ventricular tachycardia (VT) and aystole partly rests on the power spectrum of the ECG (Nygards and Hulting 1977).

The general configuration of the system is shown in Figure 2. A bipolar ECG is obtained from standard bedside equipment. Monitoring status and arrhythmia diagnoses are presented on a video terminal in the nurses station, as well as on slave monitors in the patient rooms. A total of 17 arrhythmia diagnoses can be made by the system. The arrhythmias and their definitions are listed in Table I.

Table I.   Arrhythmia Alarms

Priority	Message	Explanation
***	ASYSTOLE	R-R > 5.0 s and power below threshold.
***	VENTRICULAR FIBRILLATION	Abnormal pattern for 5 s and peak in power spectrum.
***	VENTRICULAR TACHYCARDIA	More than 3 consecutive VBs and rate > 120/min or narrow peak in power spectrum.
**	RUN OF MORE THAN 3 VB'S	More than 3 consecutive VBs with rate ≤120/min.
**	IDIOVENTRICULAR RHYTHM	More than 3 consecutive VBs with rate ≤40/min or QRS width > 0.14 s and HR ≤ 40/min.
**	BRADYCARDIA	HR more than 15 beats/min below bradycardia limit.
**	TACHYCARDIA	HR more than 30 beats/min above tachycardia limit.
**	RUN OF 2-3 VB'S	2 or 3 consecutive VBs.
**	R-ON-T VB'S	2 early VBs of the same shape within 15 min.
**	VENTRICULAR BIGEMINY	3 VBs alternating with normal beats.
**	SUPRAVENTRICULAR TACHYCARDIA	More than 3 consecutive SVBs with a rate above tachycardia limit.
*	MORE THAN 5 VB'S/MIN	
*	MULTIFORM VB'S	2 differently shaped VBs within 5 min.
*	MISSING QRS	R-R interval between normal beats at least 50% prolonged on two occasions during the past miN and basic rhythm regular.
*	MORE THAN 5 SVB'S/MIN	
*	BRADYCARDIA	HR below bradycardia limit (normally 50/min).
*	TACHYCARDIA	HR above tachycardia limit (normally 120/min).

Reproduced from Acta Med Scand Suppl 630 with permission

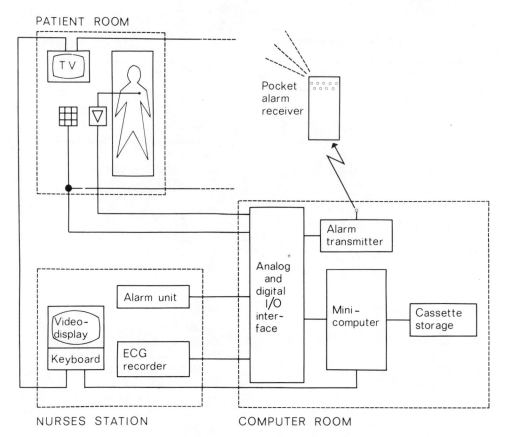

PATIENT ROOM

Pocket
alarm
receiver

Alarm
transmitter

Analog
and
digital
I/O
inter-
face

Alarm unit

Video-
display

Keyboard

ECG
recorder

Mini-
computer

Cassette
storage

NURSES STATION   COMPUTER ROOM

Fig. 2 Configuration of the arrhythmia monitoring system.
(Reproduced from Acta Med Scand, Suppl 630, with permission)

II. Alarm Presentation

Each arrhythmia, detected by the computer, gives off an alarm
and an ECG write-out for manual confirmation. Alarms are grouped
in three priority levels according to the gravity of the arrhythmia. The priority levels are shown in Table I. Alarms are presented to the staff in the form of acoustical signals via wireless
pocket receivers, as well as in the form of light signals, bedside
and in the central station (Table II). The wireless alarm system
was incorporateded in order to enable the nurse to move around
freely with little risk of over looking an essential arrhythmia.
Alarms can be reset from bedside or the central station, but automatic reset is performed after a certain time (usually 4 min) if
the alarm condition is no longer present.

Table II.   Alarm presentation

Priority	Visual (alarm panel)	Acoustic (pocket receiver)	Video Screen
Top (XXX)	Flashing red light	Repetitive sounds	Arrhythmia
Middle (XX)	Steady light	Two short sounds	Arrhythmia diagnosis
Low (X)	-	-	Arrhythmia diagnosis

III.   Evaluation of ECG monitoring

The performance of the system has been evaluated clinically in two studies.   In the first, a continuous ECG was recorded for ten 24-hour periods in the ICU (1000 monitoring hours) and was used   as a reference (Hulting and Nygårds 1978). The computer write-outs were compared with this reference. About 90% of the patients with various ventricular and supraventricular arrhythmias were detected by the automated system (Hulting 1979b).

In the second study the performance of the system with refer- ence to the highest priority alarms (asystole, VF and VT) was eva- luated during a 4-week period (Hulting 1979a). In five patients a total of 24 asystole episodes were detected manually. Eight of these were accompanied by a correct alarm and two remained totally undetected by the computer. Four episodes of VF were found in four patients. The computer diagnosis was correct in three and VT in one patient.   Thirty-one out of 44 runs of VT were correctly de- tected. Eight VT episodes were completely missed by the automated system.

IV.   Arrhythmia spectrum in the ICU

All ECG write-outs from the automated system have been col- lected during one year of clinical use (Hulting 1979c). During this period 679 patients were admitted to the   ICU   with   suspected AMI.   This group resolved in 339 patients with proven AMI and 340 patients without proven AMI. All studied arrhythmias, except atri- al fibrillation of flutter and bradycardia were significantly more common in the AMI group (Table III). In patients with AMI, 4   ar- rhythmias were associated with a significant increase in mortality, namely asystole, extreme bradycardia, VF and tachycardia. Patients in the AMI group with paired VBs and SVT had a significantly lower

ICU mortality than those without these arrhythmias. In the other arrhythmias, including VT, no prognostic conclusions regarding mortality could be made. With the use of more restricted definitions of VT, however, it was found that AMI patients with a high rate (150/min) or longer runs (7 complexes) of VT had a statistically significant increase in mortality (Hulting 1979b). These observations have resulted in some modifications regarding our policy for antiarrhythmic treatment.

Table III. Arrhythmia incidence (%) in 339 patients with proven AMI and in 340 patients without proven AMI. (n.s. = not significant)

	AMI	Non-AMI	Difference
Aystole	7	2	$p < 0.01$
VF	6	2	$p < 0.05$
VT	38	10	$p < 0.001$
Run of VBs	50	17	$p < 0.001$
Extreme bradycardia	11	6	n.s.
Paired VBs	72	38	$p < 0.001$
R-on-T VBs	15	7	$p < 0.01$
Ventricular bigeminy	31	20	$p < 0.01$
SVT	50	22	$p < 0.001$
More than 5 VBs/min	49	29	$p < 0.001$
More than 5 SVBs/min	45	28	$p < 0.001$
Bradycardia	39	45	n.s.
Tachycardia	55	34	$p < 0.001$

Reproduced from Acta Med Scand Suppl 630 with permission

V. General information system

The experience gained from our own projects and from the study of a number of units in Europe and North America have been used in the planning of a new 24-bed ICU. This unit has been built up rather unconventionally. Some basic features are:

- Ward design based on complete computerization
- Physical resources in surplus
- Normally 24 beds but up to 32 if necessary
- 1-2 patient rooms only
- All rooms self-supporting with nursing accessories

With regard to computerization, the following prerequisites were formulated:

- Automatic non-invasive measurements of cardiopulmonary vital parameters
- All patient information included (except for case history)
- Wireless paging system at alarms
- Subsystems operated through a few function keys and a numeric keyboard at one type of terminal distributed in all rooms
- Decentralized access to all information
- Independent back up system for optimal safety

The information system has been designed by the Mennen-Greatbatch Co. in close cooperation with the staff of the ward.

VI. Hardware Configuration

The information system is based on four PDP 11 T 34 minicomputers (Fig. 3). Two of these exclusively deal with arrhythmia monitoring. One processor is reserved for back up function in case of failure in one of the others. Computer terminals (32) are dis-

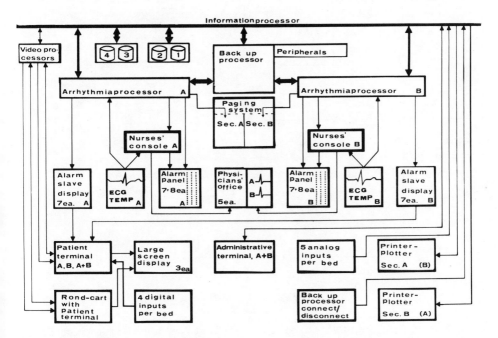

Fig. 3 Hardware configuration of the general information system

tributed all over the ward, making possible a fully decentralized use of the system with minimal interference with ordinary patient care routines. Data are fed into the system either manually from keyboards (e.g. fluid balance, medication, blood gas values, doctors' and nurses' notes) or served on-line (e.g. blood pressure, temperature, heart rate).

VII.   Respiratory Monitoring

For the monitoring of respiratory parameters a quadrupole masspectrometer (Centronic 200 MGA) will be attached to the information system. Respiratory gases will be sampled at bedside and conducted by a capillary tubing system to a centrally positioned masspectrometer. The sampling sequences and the preprocessing of respiratory data as well as calibration of the masspectrometer will be performed by a separate microprocessor (Fig. 4).

After complete built out of the system we anticipate continuous monitoring of circulatory and respiratory vital parameters in all critically ill patients.

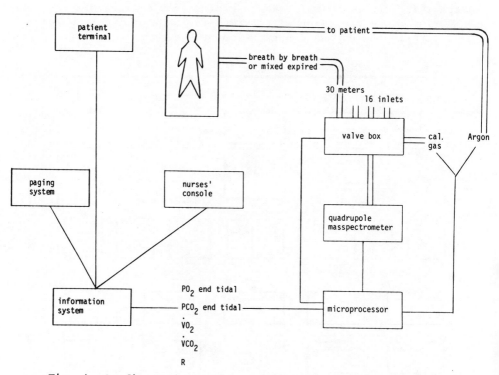

Fig. 4   Configuration of the respiratory monitoring system

The first patients were admitted to the new ward in early May 1979. At present, only the information system and the arrhythmia system are operative. The respiratory monitoring will be ready for use in October 1979.

REFERENCES

1. BEVEGÅRD, S., HALLDIN, M., HULTING, J., MALMLUND, H-O. et al: Computer control of blood pressure in cases of severe barbiturate poisoning (Swedish). Report from The Department of Aviation and Naval Medicine, Karolinska Institutet, Stockholm, Aug 1969.

2. HULTING, J. and NYGÅRDS, M-E.: Accuracy of alarms from a computer-based arrhythmia monitoring system. Acta Med Scand 203: 153-159, 1978.

3. HULTING, J.: Detection of asystole, ventricular fibrillation and ventricular tachycardia with automated ECG monitoring. Acta Med Scand 205:17, 1979a.

4. HULTING, J.: Application and evaluation of automated arrhythmia monitoring in the coronary care unit. Acta Med Scand (Suppl) 630, 1979b.

5. HULTING, J.: Arrhythmias in the CCU recognized with the aid of automated ECG monitoring. Acta Med Scan (accepted for publication) 1979c.

6. NYGÅRDS, M-E., TRANESJÖ, J., ATTERHÖG, J-H. et al: On-line computer processing of pressure data from cardiac catheterizations. Comp Progr Biomed 5:272, 1976.

7. NYGÅRDS, M-E. and HULTING, J.: Recognition of ventricular fibrillation utilizing the power spectrum of the ECG. Computers in Cardiology, p. 393, IEEE Computer Society, 1977.

8. NYGÅRDS, M-E. and HULTING, J.: An automated system for ECG monitoring. Comp Biomed Res (accepted for publication), 1979.

MASS SPECTROMETRY IN ANAESTHESIA

A. A. Spence, W. O. M. Davis,
R. Allison, and H. Y. Wishart

University Department of Anaesthesia
Western Infirmary
Glasgow, Scotland

Although mass spectrometers have an established role in analysis of the normal respiratory gases, their use in the presence of anaesthetic gases has been limited and often unsuccessful. The major problems have been the interference of fragmentation components of molecules such as nitrous oxide and halothane with the signal from nitrogen, oxygen and carbon dioxide.

As the ratio between fragmentation peaks is constant for any one gas or vapor for a period of at least 24 hours, circuitry was developed into which this ratio could be set for the MGA 200 quadrupole apparatus (Centronic Limited, Croydon, England). Circuitry to add the undesired fragmentation peaks and subtract them from the apparent, but erroneous, reading allows the true value of the gas to be analysed. The outputs of this spectrum overlap eraser (SOE) can be fed to the MGA 200 automatic stability control mode (ASC) circuit providing stability of calibration of +-1% of the true value of each gas or vapour for more than 24 hours.

Such a facility could have considerable application in the practice of anaesthesia:

1. Monitoring the gas composition of closed circle breathing systems.

2. Measurement of oxygen consumption and carbon dioxide output, tracer gas techniques for assessment of ventilation/perfusion relationship in the lungs, cardiac output and pulmonary interstitial water content.

3. Studies of the uptake and elimination of anaesthetic gases.

In our own hospital we have demonstrated with the mass spectrometer satisfactory conditions of anesthesia for adults using a closed circle breathing system with a fresh gas input (halothane in pre-mixed 50% nitrous oxide in oxygen; FIO2 in the range 0.3 - 0.4; spontaneous or IPPB respiratory mode) within the range 300 to 800 ml - a tenfold decrease compared with conventional U.K. practice.

The high cost of mass spectrometers and the need for skilled technical support in their use make it unreasonable to propose their widespread use unless one instrument can be shared between several anaesthetising locations. Our experience suggests that sampling over distances up to 40 metres using a nylon conduit of 0.75 mm i.d. with a delay time of 28 sec (sample flow rate 25 ml min -1) is possible without substantial distortion of the respiratory waveform. A system of time-shared sampling at present under development in the Western Infirmary of Glasgow will allow analysis of two breaths per minute from four patients simultaneously.

COMPUTERIZED ICU DATA MANAGEMENT +

John E. Brimm, M.D., Clifford M. Janson,
Richard M. Peters, M.D., Michael M. Stern, Ph.D. **

Department of Surgery/Cardiothoracic
University of California Medical Center
San Diego, California

** Hewlett Packard Co.,
Waltham, Massachusetts

+Supported by USPHS Grants GM 17284 and GM 25955

INTRODUCTION

Much effort has been spent in developing ICU computer systems to provide crisis alarms for events such as cardiac arrest and disconnection from a respirator. With the exception of arrhythmia detection in CCU care, however, alarms based on vital signs have not had significant success. This is in large part because alarms from a central computer duplicate those from bedside hardware.

Appropriate alarms from a central ICU computer are essential; however, they should have a different focus. They need to deal with interpretation and treatment failures. Examples of these more complex alarm conditions are failing to recognize cardiac tamponade, ignoring a low potassium level in the presence of alkalosis, using toxic FIO2 levels to treat low PaO2, and neglecting a low cardiac output when using continuous positive airway pressure.

In the ICU, the emphasis of treatment is to optimize vital organ function to permit recovery from acute illness or injury. Decisions are procedure oriented: Should we do something? If we do, does it move the system in the desired direction? This type of decision leads to three major types of errors: (1) failure to recognize an indication for action; (2) inappropriate action for the recognized abnormality or abnormalities; (3) failure to appreciate the effect of the action on systems other than the primary object of therapy. The ICU computer should minimize the frequency and impact of these errors.

The capacity for a computer to reach an appropriate decision requires a rich data base, and to date a significant portion of research and development effort in ICU automation has been directed

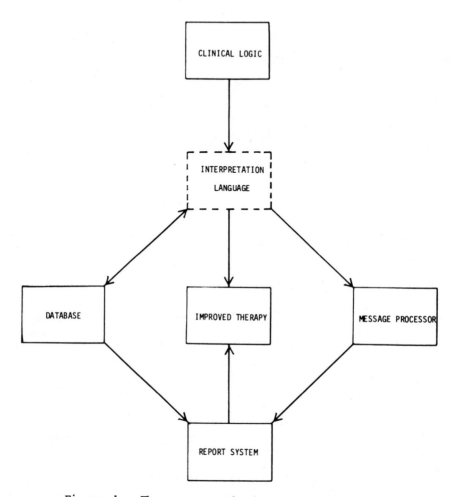

Figure 1.   The conceptual elements needed for more sophisticated alarms.

toward facile data collection.   In our opinion, we have reached the stage where the data base in certain problem domains is sufficient to allow us to develop the next stage of sophisticated alarms. Obviously, these alarms represent a different concept;   they are really interpretations of data and suggestions for treatment.

Conceptual Elements

The conceptual elements needed to provide these more sophisticated alarms are illustrated in Fig. 1. A fundamental element is medical knowledge. Without suitable knowledge, all attempts to develop computerized data interpretation are futile. Suitable problem domains must meet four criteria: (1) mismanagement must occur using existing systems (i.e., the computer system must fill a need); (2) the fundamental information needed for decision making must be able to be captured by the computer, hopefully without requiring nurses or physicians to enter it; (3) the physiological basis for decision making must be well understood; (4) the outcomes of computerized decision algorithms must be testable. Cardiopulmonary management, a primary focus of the ICU, qualifies as a suitable domain, and some of our ideas for the medical content of algorithms for cardiopulmonary management have been described in a companion paper. (1)

As depicted in Fig. 1, however, identification of the medical knowledge base is not sufficient to result in improved patient care. Two further problems remain. The first is to provide flexible and powerful software systems so that the underlying medical knowledge can be easily programmed into the computer. The second issue limiting implementation and acceptance of computerized data interpretation lies at the man/machine interface. The system elements must be coordinated to result in a product usable by physicians and nurses. Even if the computer system exceeded the capabilities of clinicians, if it were not acceptable to them, it would not be used. In order to make the system useful to and usable by the clinical staff, an organizing principle of our philosophy is that physicians should not be responsible for data entry and that output of the system should be selective and specific. In order to use the computer to aid in providing improved treatment, four key systems elements must be provided to transfer the medical knowledge base acceptably to the clinical staff -- a rich database on which to base interpretations; an interpretation language which permits concise, modular, understandable, and modifiable expression of the underlying medical logic; a message processor for managing output of the clinical logic modules and for notifying users; and a report system for presenting data from the database to justify the findings of the logic modules. These four elements - database, interpretation language, message processor, and report generator - are all required for a workable clinical system.

Database

In the past we have concentrated on providing facile methods for collecting a medically rich database. Elements of this database have been described in companion papers. (2-4) Our ICU system, which is based in the Hewlett-Packard 5600A patient data management system, has been developed specifically to acquire fundamental physiologic data regarding respiratory, cardiovascular, renal, and metabolic function without requiring the physician or the nurse to enter data: (a) vital signs are sampled directly from bedside monitors; (b) arterial and venous blood gas values are entered directly from the blood gas laboratories; (c) clinical chemistry data will be transferred directly from the main hospital computer; (d) cardiac output determinations are performed using an on-line program either by nurses or housestaff; (e) tests of pulmonary function--including functional residual capacity (FRC), analysis of intermittent mandatory ventilation (IMV), gas exchange, and lung volumes and flow rates--are done on ICU patients by technicians using on-line, mobile respiratory carts; (f) fluid and electrolyte balance data can be entered by nurses using a newly developed program.

Without this relatively complete data set, interpretation would be impossible because key information would be lacking. Thus, provision of a rich patient information set has been an essential preliminary.

The AID Interpretation Language

Development of computer programs to implement medical decision algorithms using currently available programming languages is extremely time consuming. Our experience in developing the blood gas information system(2) using FORTRAN has shown the need for a special-purpose decision language particularly with the capability of dealing with missing data. The blood gas system is a prototype for implementing decision logic. The logic took a long time to program initially, and worse, took an undue effort to revise as medical requirements changed. This experience in fact led directly to the formulation of the AID language.

The language for Assistance In Decision making (AID) is seen as a tool, not an end. AID will be able to rapidly express medical decision logic, our ultimate goal. The principal advantages of developing a special-purpose language are that: (a) its primitives will be precisely those needed to implement modular sequential decision trees; (b) it will have explicit structures for coping with missing data; (c) it will have high-level constructs for retrieving and storing data in a patient's time-structured computer record; (d) its features should permit concise, simple, understandable, and modifiable expression of the underlying medical logic. AID will be, therefore, a practical tool for the implementation of decision logic. Because the language is still under development, detailed explanation of its features is premature.

Message Processor

A vexing limitation of the computerized ICU medical record is that users cannot easily browse through it to ascertain crucial new information. For computerized outpatient records with low data rates, users can scan the record quickly. In the ICU, however, massive amounts of data are collected over short time spans, patients commonly have multiple problems, and many different people -- primary physicians, consultants, nurses, and paramedical personnel -- interact with the record. This problem is compounded by the fact that much data can now be entered into the patient's record without direct intervention on the part of the ICU staff; for example, the clinical chemistry and blood gas laboratories can store data directly and the computer can sample vital signs automatically from bedside monitors.

A partial solution is to structure retrieval from the patient's record around the patient's current problems, i.e., to use the problem-oriented structure of Weed. (5) A complementary solution is to structure retrieval by the time and priority of the information in the record, that is, to notify users about new data and interpretations.

We have developed a Message Processor which is a generalization of the graded alarms used for arrhythmia monitoring. (6) Messages are short text phrases (e.g., "SEVERE METABOLIC ACIDOSIS") with the following associated properties: class (e.g., acid-base status), time, priority, lifetime, report number, and consumer. When information is generated which activates a message for a patient, the following steps occur: (1) a line is drawn around the periphery of all CRT's in the ICU for that patient; this line is drawn and erased (i.e., "flashed") at shorter intervals as the priority of the message increases. This line-drawing technique is used because it is compatible with our current equipment but could readily be adapted to other equipment. Our method does not interfere with screen contents in case a terminal is being used. (2) When ICU personnel note the flashing screen, they may quickly choose a display of all currently active or unreviewed messages for patients in their ICU, ranked by priority, patient and time. (3) They may select a more detailed report of the message: for example, in the case of the metabolic acidosis message, the blood gases will be displayed. As users view a report, the system acknowledges their receipt of the message by removing it from the list.

A concept embedded in the system is that if a message of a particular class is posted, and a second message of the same class but higher priority is posted for the same time, then the second message replaces the first. Thus, if a logic algorithm reached a particular conclusion regarding oxygenation (the "class"), for example, to raise the FIO2, but then after considering the cardiac output, concluded that a better maneuver would be to raise the CPAP level, this second message would be the one posted to users. This feature of messages permits intermediate conclusions to be posted and provides a method for dealing with conflicts.

The Message Processor already provides for posting "future" messages. Most messages are posted to users at the time that they are received, but future messages are suppressed until an indicated time. This highly useful property permits clinical algorithms, for example, to interpret blood gases, to immediately display a message such as "METABOLIC ALKALOSIS: DECREASE THE VENTILATOR RATE", and simultaneously to post a second message to appear one hour in the future stating "DRAW ARTERIAL BLOOD GASES".

The capabilities of the Message Processor are currently being broadened. The present system does not store messages into a patient's file, so that users cannot get a display or printed list of past messages. Messages are currently used for communicating findings to users on a real-time basis only, such as to inform ICU staff that blood gas data have been entered by the lab. By storing the messages, they can be retrieved and used as a time/priority index for accessing a patient's record.

Report System

The microprocessor revolution has touched far more than just bedside instruments in the ICU. Computer peripherals now commonly include microprocessors, resulting in far more capable devices. Display terminals and line printers, the main interfaces between ICU computer systems and clinical users, now have many more features with little, if any, added cost, and prices are dropping. Display terminals now come with inverse video, blinking, split screens, double- and half-size characters, upper and lower case, multiple type fonts, different intensities, graphics, higher resolution, color, and touch-sensitive screens. Printers come with an equally bewildering set of features. Since the user interface has been a particular weakness of computer systems, these features are vital for improved human engineering.

A criticism of ICU computer systems has been that they do not "significantly alter nursing activities". (7) Indeed, they may add to the nurse's burden if double charting is required; i.e., if the nurse must keep a handwritten chart and enter the same data into the computer. Why would double charting persist in the presence of more capable output devices?

A partial reason is lack of reliability. However, if suitable hard copy is available for back-up when on-line terminals are down, lack of reliability is only a partial culprit. (If is also likely to play less of a role as fault-tolerant systems evolve.) A more basic answer is that we have not had software capable of using the features of the devices. At this writing, the Hewlett-Packard system still has no general capability for printed output, let alone the capacity to support the exotic features of newer printers and display devices. On the other hand, Hewlett-Packard has provided a flexible set of programs facilitating data entry. These generalized programs use table-drive display "forms" or "frames" with standardized conventions for users' interactions. While the programs provide an adaptable mechanism for collecting data, however, they store data in a rigid organization suitable for collection, but not for medical use.

The example of data required for fluid management should illustrate this point. Physicians make decisions regarding fluid management based on the following data: (1) fluid balance data input by nurses; (2) acid-base status based on blood gas values determined in the blood gas laboratory; (3) serum and urine electrolytes determined in the chemistry laboratory; (4) hemoglobin/hematocrit done in the hematology laboratory; (5) cardiac output performed using our on-line testing system; and (6) blood pressures sampled from bedside monitors. Most of these data are collected at random time intervals throughout the day.

The inability to easily integrate data from many different sources onto one report has inhibited use of the system. A hard-coded program could be written to merge data onto a report, but then as medical requirements changed over time, tremendous effort would have to be expended to revise the program. A much better solution is to provide a report program which can assimilate data input from various sources and stored at different times. Such a program would uncouple data entry from data reporting.

We have designed the specifications for, and have essentially completed programming, a high-level language with powerful constructs that permit users quickly and easily to tailor complex medical reports to their needs. This Report Definition Language (RDL) transparently supports different output devices including printers and display terminals, and provides for both horizontally and vertically structured tabular reports. The language design allows for enhancement with addition of graphic and text presentations. A strength of the language is its flexibility. Few restrictions are placed in intermixing data from various sources, such as the clinical laboratory, nursing entry, or on-line monitors, so that medical interpretive needs can be divorced from data source. Further, because measurements in the ICU are not made simultaneously, the system tolerates some predefined "variance" in the times at which data may be presented together. Since the system is implemented as a language, it permits new reports to be easily modified. The language is defined with a precise syntax and "application" report programs are self-documenting. Continuing development of this language, particularly to support new device features, is essential.

Provision of the RDL language is a tool and the tool must still be integrated into the system. In particular, we must provide flexible scheduling programs for invoking the report programs. As we described in a review of ICU data handling systems, (8) ICU systems have a capability that has never been exploited. Because the data capture and reporting functions are separate, computer-generated reports can conform to the level of sophistication and need of the user. Manual systems require all users to gather data from the same record, and to date, so have computer systems.

However, the important data for caring for a head injury patient may be quite different from that for a post-CABG patient. Dr. Smith might like to see data presented in a horizontal format while Dr. Jones prefers a vertical format. A respiratory therapist might need to see respiratory data presented in quite a different way from the Chief of Pulmonary Medicine. People's needs and idiosyncracies differ, and the purpose of the Report System is to be able to accommodate them easily. Thus, an important feature must be the provision of adaptable scheduling mechanisms for both supporting clinical needs and enhancing user acceptance.

Integration of the Basic Elements

Physicians and other clinical staff err when they fail to appreciate the implications of available data through oversight, fatigue, or ignorance. A system which minimized these errors by interpreting collected data would be highly beneficial. This requirement suggests that interpretation should proceed automatically as data are input to the system. A basic contention is that the ICU staff is already subject to information overload. A system which functioned only when users requested it would reinforce human limitations. Rather, the system must constantly filter all incoming data to warn users of potential problems. Naturally, users should also be able to invoke the system on demand, but this is of secondary importance. Without the onerous requirement for data entry, physicians and other ICU staff can become the beneficiaries of "computer consulting" by coupling consultation programs to the patient's ICU computer record.

The advice, or interpretations, generated by decision algorithms should become the primary reports of the ICU system. Currently, ICU computer systems typically report data in dense tabular or graphic formats. So much data is collected, much of which may be highly sophisticated, that users have difficulty synthesizing it into a coherent whole. Users are confronted with multiple pages of data with scarcely a word about what it might mean. Many users are inundated by the unimportant while others are intimidated by the sheer volume. Potentially useful, even vital, information may go unused.

We propose to print only a short interpretation message about each problem. This message could be followed up, if the user desired, with the reason for the interpretaion and a more detailed data presentation, either graphic or tabular, using the Report System. These capabilities are essential for gaining user acceptance and for permitting physicians to make judgement decisions contrary to the computer. Emphasis on presenting the most important findings first would enable users to be more selective in following up results without being overwhelmed by them.

Interpretation messages can be used to lead the user to the pertinent information in the computerized medical record. Because ICU systems capture data directly from many sources -- blood gas technicians, the clinical chemistry computer system, and automated bedside instruments -- potentially important information could be entered into an ICU system without the ICU staff's awareness of its presence or importance. Once the interpretation system has been developed, it can provide an index back to this "hidden" record. Since a property of each interpretation is a priority ranking (a measure of the significance of the interpretation), the time/priority oriented medical record is perhaps a more useful, or at least a complementary, organization for acute care than Weed's (5) problem-oriented structure. Neither problem-orientation nor time/priority-orientation changes the underlying database; they are merely alternative methods of access.

Thus, to return to the conceptual elements illustrated in Fig. 1, the AID language is seen as the key element for implementing medical logic. AID modules will filter data coming into the system, and will store conclusions back into the patient's database. Significant findings will be broadcast to clinical users via the Message Processor. Messages can be followed up in more detail using the Report System, which draws on the database.

Of these four elements, the database, Message Processor, and the Report System are essentially complete although they will always be evolving. Work on the AID language is underway, and integration of all the elements awaits its completion.

REFERENCES

1.   PETERS, R.M., J.E. BRIMM and C.M. JANSON. Clinical basis and
     use of an automated ICU testing system. Proc. First Annual
     International Symposium Computers in Critical Care and Pulmonary
     Medicine, Norwalk, Connecticut, May 24-26, 1979.

2.   BRIMM, J.E., C.M. JANSON, M.A. KNIGHT and R.M. PETERS. Extract-
     ing information from blood gas analyses. Proc. First Annual
     International Symposium Computers in Critical Care and Pulmonary
     Medicine, Norwalk, Connecticut, May 24-26, 1979.

3.   BRIMM, J.E., M. HALLORAN, C.M. JANSON and R.M. PETERS. Computer
     based fluid balance records. Proc. First Annual International
     Symposium Computers in Critical Care and Pulmonary Medicine,
     Norwalk, Connecticut, May 24-26, 1979.

4.   JANSON, C.M., J.E. BRIMM and R.M. PETERS. Instrumentation and
     automation of an ICU respiratory testing system. Proc. First
     Annual International Symposium Computers in Critical Care and
     Pulmonary Medicine, Norwalk, Connecticut, May 24-26, 1979.

5.   WEED, L.L. Medical Records, Medical Education, and Patient Care:
     The Problem-Oriented Record as a Basic Tool. Cleveland: Case
     Western Reserve University Press, 1969.

6.   SANDERS, W.J., E.L. ALDERMAN and D.C. HARRISON. Alarm processing
     in a computerized patient monitoring system. In Computers in
     Cardiology, 1975, pp. 21-25.

7.   EDMUNDS, L.J., JR., H. MCVAUGH, ILL, and J. STEVENS et al.
     Evaluation of computer-aided monitoring of patients after heart
     surgery. J. Thorac. Cardiovasc. Surg. 74: 890, 1977.

8.   BRIMM, J.E. and R.M. PETERS. Data handling systems. In J.M.
     Kinney (ED.), Manual of Surgical Intensive Care of the American
     College of Surgeons. Philadelphia: Saunders, 1977, Ch. 7.

A COMPUTER ACTIVATED TWO-VARIABLE TREND ALARM

A.M.   Benis, H.L.   Fitzkee
R.A.   Jurado, R.S.   Litwak

Division of Cardiothoracic Surgery
Mt. Sinai Medical Center
New York, New York

In our cardiac surgical intensive care unit the cardiovascular status of patients having undergone cardiac surgery is assessed with the aid of the monitoring of mean arterial and left atrial pressures (MAP, LAP). MAP and LAP data are continuously provided at the bedside with the use of pressure transducers and digital displays (Hewlett-Packard System). Associated visual and audible alarms trigger within a few seconds if either MAP or LAP stray outside of their preset limits. In addition, a central computer (IBM 1800) routinely samples the MAP and LAP every 10 minutes. The computer has been programmed to activate an alarm sequence if a routinely acquired value of MAP or LAP is found to be outside of preset ("trend") limits. The computer then initiates a repeat mode of sampling of MAP and LAP at 1 minute intervals. If 3 consecutive repeated values of either variable remain outside of the preset limits then the following occur: 1) A plot of the trend of MAP and LAP over the preceeding 3 hours appears on the bedside video monitor, 2) A bedside light and audible chime are activated, and 3) The bedside keyboard is locked out to other functions until the alarm is reset by a qualified staff member with the use of a personalized code. A pilot study showed that this system could detect adverse trends with excellent reliability, averaging only 1-2 false alarms per patient during an 8-hour shift.

INTRODUCTION

As monitoring of patients in the intensive care unit has become more complex, it has become increasingly more desirable to utilize the computer in an alarm function. Some recent sophisticated statistical monitoring techniques have been reviewed by Lewis (1), Hope and coworkers (2), and Endresen and Hill (3). Such techniques have the potential advantage of forecasting adverse conditions before they occur, but have the disadvantage of complexity and thus have not yet found routine application.

Alarm systems that are activated when upper or lower limits of a variable are exceeded are notoriously prone to false alarms. Raison and coworkers (4) found that alarms on the limits of heart rate (EKG signal) and arterial pressure were mostly spurious. They concluded that it was more practical to trigger the alarm only when coincident changes in both heart rate and arterial pressure occurred.

The purpose of the present study was to develop a simple alarm system that would detect adverse trends in mean arterial pressure (MAP) and left atrial pressure (LAP), with a minimal incidence of false alarms, in post cardiac surgery patients.

Alarm System

Basic hypotheses

In this study untoward clinical events are divided into two classes:

    I Acute Episode
    II Adverse Trend Condition

Events of the first class, Acute Episode, are defined as those that occur on a time scale of 0-15 minutes. Those of the second class, Adverse Trend Condition (ATC), are events that occur more insidiously, on a time scale of 15 minutes or longer. The present study is confined to the monitoring of MAP and LAP, hence the relevant ATC is defined as an alteration in MAP or LAP over a time of 15 minutes or greater, requiring reevaluation of therapy by the physician.

The specific objective of the alarm system presented herein was to detect events of the ATC class but not necessarily to avoid detection of all Acute Episodes.

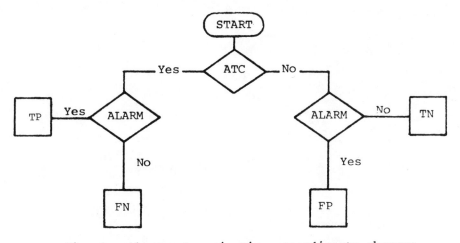

Fig. 1 - Alarm categorization according to absence or presence of ATC and alarm (TP = true positive, FN = false negative, FP = false positive, TN = true negative).

Categorization of Alarms

The four alarm categories are presented in the schema shown in Fig. 1. They are as follows:

True positive (TP) - Verified ATC with appropriate alarm.

False negative (FN) - Verified ATC with no alarm.

True negative (TN) - No ATC with no alarm.

False positive (FP) - No ATC with inappropriate alarm.

The alarms are further subcategorized as shown in Table 1. With regard to the true positive alarms, X signifies that the alarm preceded physician confirmation of an ATC, Y signifies that detection of an ATC by the staff preceded the triggering of the alarm, and Z signifies the repeated triggering of an alarm after having been reset during a continuing ATC.

False alarms are subcategorized in Table 2 according to whether the cause was due to problems with instrumentation, operator, alarm system or patient instability.

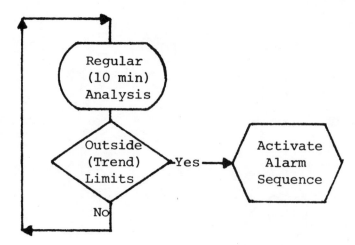

Fig. 2 - Computer alarm algorithm.

Adverse Trend Conditions (ATC's) were classified according to the clinical cause and therapy administered (Tables 3
and 4), the classification being particularly relevant to alterations in cardiovascular status of patients having undergone cardiac surgery.

ICU Setting and Alarm System

The present system was developed for patients in a cardiac surgical intensive care unit (ICU) who had undergone open
cardiac surgery (5). Patients are continually assessed with
the aid of the monitoring of mean arterial and left atrial
pressures (MAP,LAP) with the use of indwelling catheters,
pressure transducers, and bedside digital displays
(Hewlett-Packard System). Associated visual and audible
alarms trigger within a few seconds, if MAP or LAP stray outside of their preset limits.

In addition to the bedside display of MAP and LAP, a central computer (IBM 1800) routinely samples the MAP and LAP
every 10 minutes (6). If a routinely acquired value of MAP or
LAP is found to be outside of preset ("trend") limits, the
computer activates a programmed alarm sequence (Fig. 2). In
this sequence the computer initiates a repeat mode of sampling
of MAP and LAP at 1 minute intervals (Fig. 3). If three consecutive repeated values of either variable remain outside of
the preset limits then the following events occur: 1) A plot
of the trend of MAP and LAP over the preceding three hours appears on the bedside video monitor, 2) A bedside light and audible chime are activated, and 3) the bedside keyboard is

locked out to other functions until the alarm is reset by a qualified staff member with the use of a personalized code. As shown in Fig. 3, if a value of MAP or LAP previously outside of limits falls once again within limits, then the computer continues to sample at one minute intervals until either three successive values are outside the limits (alarm sounds) or three successive values are within limits (regular analysis every 10 minutes resumes).

The algorithm allows monitoring of either MAP or LAP, or both simultaneously. As presently programmed, after an alarm is detected and reset, it is inhibited for a period of 15 minutes to allow appropriate action by the staff. It is to be noted that, considering that the longest time between sampling is 10 minutes, the algorithm will permit a variable to be outside of the limits for not longer than 13 minutes before triggering the alarm. This is consistent with the objective of the alarm system, that is to detect ATC's which are by definition of 15 minutes or longer in duration.

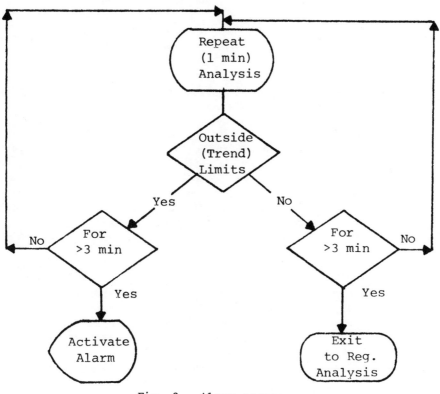

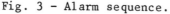

Fig. 3 - Alarm sequence.

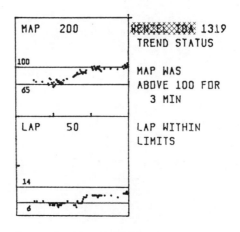

Fig. 4 - Plot of trend of MAP and LAP for typical
patient as seen on video monitor. Abscissa is time in
hours. Full scale ordinate values are 200mm Hg for MAP
and 50mm Hg for LAP. The plot shows the appropriate
detection of an ATC due to a progressive trend in MAP.

Fig. 4 shows an example of a display of the trend of MAP and
LAP during an ATC for a typical patient, together with typical limits for these variables.

According to the above algorithm, the alarm will be activated
when a variable, previously within acceptable limits, is subsequently found on repeated sampling to be outside of these limits.
Considering Webster's definition of "trend" (7) as "the general
movement in the course of time of a statistically detectable
change", it is felt appropriate to call the system a "trend alarm"
and the limits "trend detection limits".

Evaluation of Alarm System

Assuming observations made at 15 minute intervals, in accord
with both the definition of an ATC and the sampling frequency of
the alarm system, each observation can be categorized as True Positive, True negative, False positive or False negative. Using the
standard terminology for the evaluation of a test with respect to
its ability to detect a given condition (8,9), the various indices
characterizing the alarm system are shown in Table 5. These have
the following significance:

Sensitivity - A poor sensitivity (significantly less than 1.0) would imply a high rate of false negatives, i.e. ATC's missed by the alarm system.

Specificity - A poor specificity ($\ll 1.0$) would imply that triggering of the alarm is not specific to ATC's, being responsive to other spurious causes, i.e. prone to false alarms.

Predictive Value of Positive Test - This denotes the proportion of all alarms that are appropriate.

Predictive Value of Negative Test - This denotes the proportion of silent intervals that are appropriately silent.

Prevalence of ATC's - This denotes the fraction of intervals in which an ATC is detected. This parameter is largely determined by the patient population.

Incidence of Alarms - This quantity is expressed as number alarms per patient per 8-hour shift, based on observations made every 15 minutes.

Incidence of False Alarms - This quantity is termed the "Annoyance" and is also expressed per patient per 8-hour shift. The time period of 8 hours is chosen because it is mainly the nurse at the bedside to whom false alarms will be an annoyance.

Alarm Reliability - This is the proportion of intervals in which the alarm system is operating appropriately.

It was anticipated, before testing the alarm system, that the prevalence of ATC's would be relatively low with a correspondingly high incidence of TN'S. Hence the indices having TN in both numerator and denominator would be relatively high (> 0.9). These include Specificity, Predictive Value of Negative Test, and Alarm Reliability. Thus, the most relevant indices for evaluation of the alarm system were anticipated to be the Sensitivity, Predictive Value of Positive Test, Incidence of Alarms, and Incidence of False Alarms.

Pilot Study

In order to evaluate the alarm system, a pilot study was carried out in the ICU without the knowledge of the staff (alarm light and chime suppressed). An observer was present in the ICU at all times during the study periods, offering the explanation that he was "testing a new computer program". Data on each patient were recorded at 15 minute intervals, including the alarm class (Acute Episode or ATC), category and subcategory (Table 1), and type (MAP or LAP). If an ATC was judged to be present, the clinical cause and therapeutic mode (Tables 3 and 4) were recorded by the observer after consultation with the nurse at the bedside. Hard copies of the type shown in Fig. 4 were obtained every two hours.

It is evident that both the sensitivity and specificity of alarm function are highly dependent on the trend limit settings chosen. The initial limits were selected on the philosophy that a trend of MAP or LAP outside of these limits should obligatorily be brought to the attention of the physician on duty. The limits should be consistent with values of MAP and/or LAP requiring institution or reevaluation of therapy. Thus, setting the limits too wide would result in the system's missing ATC's (False negatives), while setting them too narrow would result in alarms occurring in instances were no true ATC existed (False positives).

Two subgroups of patients were studied, as shown below:

SUBGROUPS OF STUDY

	OPERATIVE DAY	POSTOP DAY
NO. PATIENTS	11	12
NO. INTUBATED	11	3
HOURS	61.5	60.3

The patients were studied each for a period of four to six hours, with observations being made at 15 minute intervals (Fig. 5). In the first subgroup the alarm system was activated one hour after the patient returned to the ICU following open cardiac surgery. In the second subgroup the system was activated at about 8:00 a.m. on the first postoperative day. Note that the first subgroup consisted of intubated patients, but only three patients of 12 remained intubated during the study periods of the second subgroup. Patients were selected at random for the study provided they were confined to bed and had functioning arterial or left atrial lines. In the second subgroup the study was discontinued if the patient was mobilized from bed to chair. No study was discontinued because of non-functioning arterial or left atrial lines.

On the basis of prior experience with post-cardiac surgery patients, the following initial trend limits were proposed:

MAP 80 $\pm$ 15 mm Hg (lower/upper 65/95)

LAP 10 $\pm$ 4 mm Hg (lower/upper 6/14)

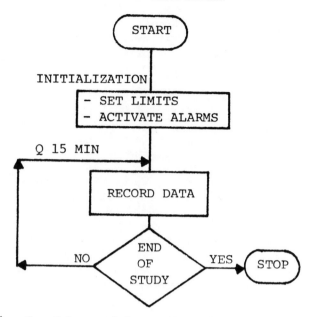

Fig. 5 - Schema of data collection for pilot study (observations made every 15 min).

These limit settings were found to be appropriate in 15 of the 23 study periods for MAP and 13 of 23 for LAP. Somewhat different initial limits for MAP were chosen in three study periods (lower limit 60 or upper limit 100) in previously hypo- or hypertensive patients. Similarly, more appropriate initial limits were chosen for LAP in five study periods, while maintaining a limit spread of +4 mm Hg around the "desired" LAP (determined on consultation with the ICU physician). After the alarms were activated it was found necessary to alter the MAP limits in five study periods, but never more than once. Similarly, it was found appropriate to alter the LAP limits once in six study periods. In only one study period of the 23 were the LAP limits changed twice.

Table 6 summarizes the results for untoward clinical events occurring during the study. There occurred two Acute Episodes each on the operative and postoperative days, with one of these episodes being (fortuitously) detected by the alarm system. These episodes involved rapid changes in MAP or LAP requiring immediate attention by the staff. On the other hand, of 33 ATC's occurring on the operative day and 10 ATC's occurring on the postoperative day, all were appropriately detected by the alarm system. That is, no False negatives were encountered. It should be emphasized that although detection of Acute Episodes was not the purpose of this system, such detection is not inappropriate.

Table 7 shows results for Specificity, Predictive Value of Negative Test and Alarm Reliability. These indices are uniformly high because of the high incidence of true negative intervals and also because of the absence of false negatives.

Table 8 shows more illuminating indices for evaluation of the alarm system. The Sensitivity of 1.00 for both days reflects the absence of False negatives. The Incidence of Alarms on the postoperative day was lower than on the operative day, reflecting a lower Prevalence of ATC's (0.13 on the operative day vs. 0.04 on the postoperative day). However, the Incidence of False Alarms (per patient per eight-hour shift) was in the range of one to two, being about the same on both days. Thus, there occurred fewer total alarms on the postoperative day, but a greater proportion of them were false. This is reflected in the Predictive Value of Positive Test of 0.75 on the operative day and 0.42 on the postoperative day. These results indicate that although the ratio of false alarms in relation to true alarms was significant, particularly on the postoperative day, the "Annoyance" or Incidence of false alarms per shift was tolerably small.

A closer examination of the false alarms is shown in Table 9 where it is seen that most of these alarms on both days were of the LAP type. Table 10 shows that on the operative day the cause was mainly patient instability (turning patient, suctioning), while on the postoperative day the cause was mainly improper leveling of the LAP transducer (often in a patient in a semi-Fowler position). Although alarms due to improper leveling of the LA transducer were placed in the FP category, these alarms are not altogether inappropriate. In fact, they bring to the attention of the staff a situation that requires remedial action.

A closer examination of the true alarms is shown in Tables 11, 12 and 13. Table 11 shows that both MAP and LAP alarms occurred to a significant extent on both days, with a relative ratio MAP/LAP alarms of about 2:1. This is in contrast to the false alarms which were mainly of the LAP type. Table 12 shows the therapeutic measures employed, with volume expansion and a vasodilator (nitroprusside) being used most frequently on the operative day, and volume expansion and a diuretic (furosemide) most frequently on the postoperative day. Table 13 shows the X, Y and Z subcategories of the true alarms, as presented in Table 1. It is of note that in a significant fraction of the true alarms (7 of 43) the alarm would have sounded before the staff's detection of the adverse trend. Of note also is a total of 11 of 43 alarms which were simply repeat alarms during extended therapy, a type of alarm that could easily be suppressed by a change in the alarm algorithm.

Improvements in Alarm System

The pilot study indicated that the Incidence of false alarms was easily in the tolerable range. However, they probably could be reduced further. False alarms (mainly of the LAP type) might be reduced by increasing the alarm trigger interval from three minutes to five minutes (Fig. 3). This would increase the maximum time that a variable could remain outside of limits without triggering the alarm from 13 to 15 minutes, but this would still be within the definition of an ATC as presented herein.

True alarms of the "Reset" subcategory (Table 1), although appropriate in their identification of an ATC, were found to be redundant. That is, during extended treatment of an adverse condition the staff was invariably well aware of the problem and would not appreciate repetitions of the alarm at 15 minute intervals. Thus, alarms in the "Reset" subcategory could be eliminated by inhibition of the alarm trigger during a continuing ATC.

CONCLUSION

This study showed that the Prevalence of adverse trends in MAP and LAP was significant on both the operative and first postoperative days, thus justifying the need for an alarm system to monitor these trends. The Sensitivity of detecting ATC's was perfect (i.e. 1.00) in the pilot study, reflecting the absence of False negatives, hence verifying the ability of the alarm to function as planned. The Incidence of false alarms was easily in the tolerable range, averaging on the order of 1-2 per patient per eight-hour shift. False alarms could probably be reduced further by increasing the alarm trigger interval, by suppressing the alarm during manipulation of the patient and by training the staff in the importance of maintaining the pressure transducers at their proper level.

The true alarms encountered were of both the MAP and LAP types (ratio of about 2:1) on both the operative and postoperative days. On both days a significant number (about 20%) of the true alarms detected by the system had not yet been detected by the staff, pointing to the potential utility of this system.

Patients of the first subgroup may be especially good candidates for monitoring of adverse trends with the system described. These intubated patients had a higher Prevalence of ATC's with a correspondingly lower proportion of false alarms. In addition, ATC's in these patients probably have a more marked life-threatening character.

TABLE 1 ALARM CATEGORY

TRUE POSITIVE

    X – Computer alarm followed by physician
        confirmation

    Y – Physician detection followed by
        computer alarm

    Z – Continuation of treatment

TRUE NEGATIVE

    N – Within normal limits at time of
        observation

FALSE POSITIVE

    A – Instrumentation

    B – Operator

    C – System

    D – Patient

FALSE NEGATIVE

    A – Instrumentation

    B – Operator

    C – System

    D – Patient

TABLE 2 CAUSE FOR FALSE ALARMS

A - INSTRUMENTATION

- Electronic failure

- Mechanical failure

B - OPERATOR

- Leveling transducers

- Calibration error

- Keyboard error

C - SYSTEM

- Algorithm

- Limit settings

- Computer

D - PATIENT

- Instability

TABLE 3 CLINICAL CAUSE

1.   ↑   Afterload
2.   ↓   Afterload
3.   ↑   Contractility
4.   ↓   Contractility
5.   ↑   Preload
6.   ↓   Preload
7.   ↓ Myocardial compliance
8.   Arrhythmia
9.   Combination
10.   Other

TABLE 4 CLINICAL THERAPY

1.   Vasodilator
2.   Vasoconstrictor
3.   ↓ Inotropic state
4.   ↑ Inotropic state
5.   Preload reduction
6.   Volume expansion
7.   Surgical procedure
8.   Antiarrhythmic agent
9.   Combination
10.   Other
11.   Observe

TABLE 5 INDICES OF CHARACTERIZATION

$$\text{Sensitivity} = \frac{TP}{TP + FN}$$

$$\text{Specificity} = \frac{TN}{TN + FP}$$

$$\text{Predictive Value of Positive Test} = \frac{TP}{TP + FP}$$

$$\text{Predictive Value of Negative Test} = \frac{TN}{TN + FN}$$

$$\text{Prevalence of ATC} = \frac{TP + FN}{TP+FN+FP+TN}$$

$$\text{Incidence* of Alarms} = \frac{TP + FP}{TP+FN+FP+TN} \times 32$$

$$\text{Incidence* of False Alarms} = \frac{FP}{TP+FN+FP+TN} \times 32$$

$$\text{Alarm Reliability} = \frac{TP + TN}{TP+FN+FP+TN}$$

---

*For 8 Hour Shift/Patient

TABLE 6 COMPUTER DETECTION
OF
UNTOWARD EVENTS

		N	DETECTED	MISSED
ACUTE EPISODE	DAY 0	2	0	2
	DAY 1	2	1	1
ADVERSE TREND CONDITION	DAY 0	33	33	0
	DAY 1	10	10	0

Day 0 = operative day

Day 1 = postoperative day

TABLE 7 INDICES OF TYPE

$$\frac{TN+\ldots}{TN+\ldots}$$

	OPERATIVE DAY	POST-OP DAY
Specificity	0.95	0.94
P.V. of Negative Test	1.00	1.00
Alarm Reliability	0.96	0.94

ERRATUM

There is a systematic error in the Table of Contents printed on pages vii–x. The corrected Table of Contents appears below.

CONTENTS

Computers in Critical Care and Pulmonary Medicine • Edited by S. Nair • 0-306-40449-4

CONTENTS

TABLE 8 INDICES FOR EVALUATION

	OPERATIVE DAY	POST-OP DAY
Sensitivity	1.00	1.00
P.V. of Positive Test	0.75	0.42
Prevalence of ATC's	0.13	0.04
Incidence* of Alarms	5.5	3.0
Incidence* of False Alarms	1.4	1.8

*For 8-hour shift/patient

TABLE 9 ANALYSIS OF FALSE ALARMS

TYPES

	OPERATIVE DAY	POST-OP DAY
MAP	0	1
LAP	11	11
MAP & LAP	0	2
TOTAL	11	14

TABLE 10 ANALYSIS OF FALSE ALARMS

CAUSES

OPERATIVE DAY	POST-OP DAY
(8) Patient, Instability	Operator, Leveling (11)
(2) Operator, Leveling	Patient, Instability (3)
(1) Operator, Calibration	

	TOTALS	
(11)		(14)

TABLE 11 ANALYSIS OF TRUE ALARMS

TYPES

	OPERATIVE DAY	POST-OP DAY
MAP	19	6
LAP	9	4
MAP & LAP	5	0
TOTAL	33	10

TABLE 12 ANALYSIS OF TRUE ALARMS

THERAPY

OPERATIVE DAY		POST-OP DAY	
(12)	Volume expansion	Volume expansion	(3)
(9)	Vasodilator	Observe	(3)
(9)	Observe	Diuretic	(2)
(2)	Inotrope	Vasodilator	(1)
(1)	Vasoconstrictor	Vasoconstrictor	(1)
(33)		TOTALS	(10)

TABLE 13 ANALYSIS OF TRUE ALARMS

	OPERATIVE DAY	POST-OP DAY
ALARM - → physician	5	2
PHYSICIAN - → alarm	19	6
RESET	9	2
TOTAL	33	10

References

1. Lewis C. D., Statistical monitoring techniques, _Med. Biol. Engng._ 1971 9 315-323.

2. Hope C.E., Lewis C.D., Perry I. R. and A. Gamble, Computed trend analysis in automated patient monitoring systems, _Brit. J. Anaesth._ 1973 45 440-449.

3. Endresen J. and D.W. Hill, The present state of trend detection and prediction in patient monitoring, _Intens Care Med._ 1977 3 15-26.

4. Raison J.C.A., Beaumont J.O., Russell J.A.C., Osborn J.J. and F. Gerobde, Alarms in an intensive care unit: an interim compromise, _Comput. Biomed. Res._ 1968 1 556-564.

5. Jurado R.A., Fitzkee H.F., DeAsla R.A., Lukban S.B., Litwak R.S. and J.J. Osborn, Reduction of unexpected life-threatening events in postoperative cardiac surgical patients: the role of computerized surveillance, _Circ._ 1977 Suppl. 56 II-44-II-49.

6. Osborn J.J., Beaumont J.O., Raison J.C.A., Russell J. and F. Gerbode, Measurement and monitoring of acutely ill patient by digital computer, _Surg._ 1965 64 1057.

7. _Webster's New Collegiate Dictionary_, Merriam, Springfield, 1976, p. 1246.

8. Barker D.J.P., _Practical Epidemiology_, Churchill Livingston, New York, 1976, p. 16-21.

9. Vecchio. T.J., Predictive value of a single diagnostic test in unselected populations, _N. Engl. J. Med._ 1966 274 1171-1173.

A COMPREHENSIVE PULMONARY DATA PROCESSING SYSTEM

Spencer L. SooHoo, Spencer K. Koerner,
Harvey V. Brown

Department of Medicine, Pulmonary Division
Cedars-Sinai Medical Center

The University of California
Los Angeles, School of Medicine
Los, Angeles, California

Acknowledgements

This work supported by grants from the Irvine Foundation, Irvine California and the Sidney Stern Memorial Trust Foundation, Los Angeles, California.

Data processing has found many uses in pulmonary medicine, including pulmonary function testing (1-3) and interpretation (4), exercise testing with gas exchange measurements (5-7), intensive care unit data acquisition and data base management (8,10). These tasks, however, are usually supported by dedicated systems or are accomplished by some type of off-line processing. Over the past year, we have developed a comprehensive system which supports all of the following tasks simultaneously:

1. On-line acquisition of blood gas data transmitted by an automated blood gas analyzer.

2. On-line acquisition of analog data in The Pulmonary Function and Exercise Laboratories.

3. Entry of patient questionnaires using an optical reader.

4. Manual Entry of Intensive Care Unit data.

We will discuss each of these applications in detail and how they are integrated into the system as a whole.

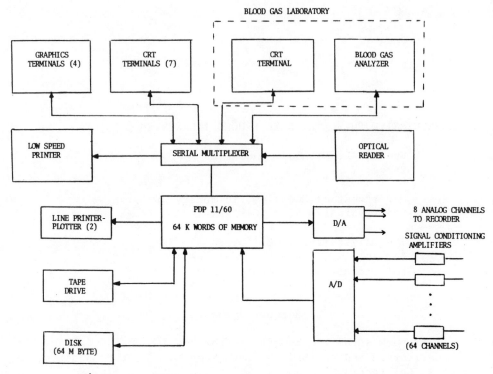

Figure 1.   Computer system hardware configuration.
The   numbers   in   parentheses   indicate the quantity of
each peripheral.

COMPUTER HARDWARE

The computer (Figure 1) is a centrally located PDP 11/60 (Digital Equipment Corporation), accompanied by a 64 megabyte disk,
tape drive, two terminals, a line printer-plotter, a 64 channel A/D
converter and 8 D/A converters.   Other peripherals located at a
distance include eight CRT terminals, four graphics terminals, an
optical reader, one low speed printer, and another printer-plotter;
the latter is interfaced to the graphics terminals for hard copy of
plots.   A Corning 175 blood gas analyzer is connected via a serial
interface.

Much of the system hardware design was dictated by the physical separation of the various tasks.   For example, analog signals
from the pulmonary function and exercise laboratories are routed to
the A/D converters over a distance of approximately 100 feet.
Another complication was the lack of uniformity in output levels
from the various instruments in the pulmonary function laboratory

(spirometers, gas analyzers, transducers, etc). These outputs ranged from 10 V to 100 mV, requiring the use of signal conditioning amplifiers in the laboratory. Signals from the testing equipment were low-pass filtered (100 Hz cutoff), amplified, and offset to provide a 20 V P-P signal for the A/D converters. The distances from the various serial devices (terminals, low speed printer, blood gas analyzer, and optical reader) to the computer ranged from a few feet to over 450 feet. This made it necessary to use 20 mA current loop instead of the more conventional RS 232 serial interfaces.

OVERVIEW OF FILES

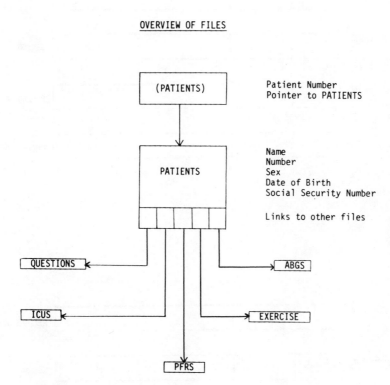

Figure 2. Relationship of the seven major data base files.

FILE STRUCTURE

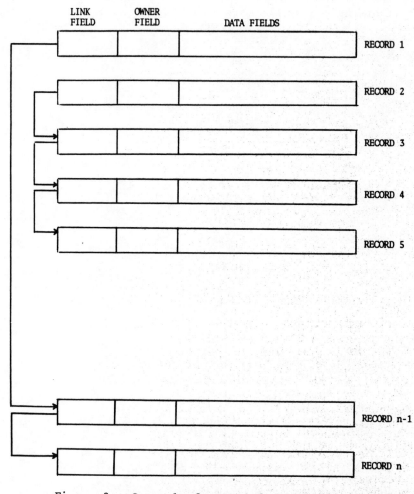

Figure 3. General format of a chained file.
Records 1,   n-1,   and n belong to one patient, and records 2, 3, and 4 belong to another patient. A -1 in
the link field indicates the end of a chain.

```
CEDARS-SINAI Division of Pulmonary Medicine              page 1 5/15/79
                         Blood Gases
    #    date time       pH  PCO2  PO2  HCO3   SO2  FIO2 Site       A-a    a/A
136884 EXAMPLE PATIENT
    1  5/15/79 1015   7.50*   43   77*  33*   96   21% 3S          19    .80
    2  5/14/79 2125   7.48*   41   78*  30*   96   21% 3S          20    .79
    3 12/05/78 0920   7.43    40   72*  26    95*  21% OB/GYN      27    .72*
```

Figure 4. Blood gas data for a single patient. The data and time indicate when the blood sample was taken, and the * denotes abnormal values. Site is an alphanumeric mnemomic corresponding to a patient location. The time, site and FIO2 are entered manually, and the pH, PCO2, PO2, HCO3, and SO2 are entered via the blood gas analyzer computer interface.

Software Considerations

    The design requirements of the computer system were that it support both an extensive data base management system and simultaneous real-time analog data acquisition from two stations in the pulmonary function laboratory and a third in the exercise laboratory. A FORTH (11,12) software package [1] was chosen over more conventional software; its advantages are that it is: 1) capable of supporting both the data base management and real-time data acquisition functioning on a single computer. 2) an interactive, structured language which facilitates the development and debugging of programs. 3) extensible, so it can be custon tailored to specific applications. 4) able to support multiple users who can access the data base files simultaneously.

    Some disadvantages that we have encountered are: 1) FORTH, as it is implemented on our system is a stand-alone, integrated language and operating system; it will not support other programming languages. 2) Many of the error checks contained in other languages are not present and have to be built into the system as required. 3) It is not a common programming language, requiring new programmers to be retrained in FORTH.

    The main file in the data base is an indirect access file which is indexed by patient number (Figure 2). The patient number (which is identical to his hospital number) corresponds to the record number in the file (PATIENTS). This file contains a pointer to the file PATIENTS, which contains patient identification data as well as pointers to other data files for the patient. These files are ABGS (arterial blood gas data), QUESTIONS (questionnaires), ICUS (intensive care unit data), PFRS (pulmonary function reports), and EXERCISE (exercise reports), all of which are chained files (Figure 3). Chaining of these files allows the data to be entered without having to reorder the file each time, and it eliminates a search of the entire file to list all of the data for an individual patient. For example, a review of a patient's arterial blood gas file involves the following sequence:

1. The user types a mnemonic command which causes the blood gas program to be compiled and loaded into his memory partition.

2. He types "find nnn" where nnn is the patient's number and also corresponds to record number nnn in the file (PATIENTS). Record nnn in (PATIENTS) contains a pointer to the patient's record in the file PATIENTS.

3. PATIENTS (Figure 2) contains -ABGS, which is a pointer to the start of the "chain" (Figure 3) for the patient's blood gas data.

4. Another mnemonic command starts a routine which follows the chained ABGS data and prints them out in the format shown in Figure 4.

5. These commands can be strung together and execute serially; in this case execution requires approximately four seconds to complete.

Although all of the major data base files (PATIENTS), PATIENTS, ABGS, ICUS, PFRS, QUESTIONS, and EXERCISE are disk resident, they are accessed as virtual memory locations.

In addition to the data base files, several temporary files exist for use as buffers for the analog data acquisition applications. As the data are sampled, they are scaled, stored in a buffer, and in the majority of the applications, plotted on a graphics terminal. After the desired information is extracted from the data, it can be overwritten. For the pulmonary function laboratory, 16 buffers, each capable of storing up to four minutes of data, are available at each of the two stations. In the exercise laboratory 20 buffers are available, each of which can store eight parameters per breath for up to 7200 total breaths.

Arterial Blood Gas Laboratory Applications

A Corning 175 blood gas analyzer with a computer interface is used in the Arterial Blood Gas Laboratory. When an analysis is complete, the results are transmitted to the computer where they are entered into the ABGS file and displayed on the CRT in the arterial blood gas laboratory. For areas such as the Respiratory

Intensive Care Unit, from which large numbers of arterial blood gases are sent, the software is configured so that the results are also transmitted to a CRT in the corresponding site. In the event of an error, there is a provision for manual data entry and editing. Blood gas data can be immediately accessed from any terminal in the system and is displayed in date and time sequence (Figure 4). The (A-a)PO2 is calculated from the alveolar air equation; a/A refers to the calculated arterial/alveolar ratio.

We are in the process of developing a microprocessor system with dual floppy disks as an intelligent terminal with the capability of communicating with both the blood gas analyzer and the PDP 11/60. If the PDP 11/60 is inoperative, the data will be retained on floppy disks for subsequent entry into the data base. This system will also provide a backup for the blood gas file portion of the data base.

Pulmonary Function Laboratory Application

In the pulmonary function laboratory, conventional instrumentation is used and the computer is configured as a "passive monitor" so that if necessary, pulmonary function testing can be performed manually. The programs for this application compute spirometry, helium dilution, single breath oxygen, single breath carbon monoxide diffusion, and body plethysmography. With the exception of plethysmography, the software is designed to acquire, store, and plot the analog data in real-time (Figure 5). After data acquistion, calculations are performed to derive the desired result. For example, in spirometry, an analog volume signal is sampled via the A/D converters, stored on disk, and plotted in realtime on a graphics CRT. The stored data can then be treated as a timed vital capacity, with calculation and display of the forced vital capacity (FVC) forced expiratory volume in one second (FEV1), FEV1/FVC, maximum expiratory flow rate (MEFR) and maximum mid-expiratory flow volume (MMFR) (Figure 6), or the maneuver can be treated as a maximum expiratory flow-volume curve (Figure 7). The timing of the forced vital capacity is by backward extrapolation from the steepest part of the volume-time tracing (13).

Up to four minutes of data can be aquired and displayed on a single trial, and up to 16 trials are allowed. For spirometry, carbon monoxide diffusion, and single breath nitrogen, data collection is ended after a time interval selected by the technician. For helium dilution, the trial is ended when two successive samples of helium taken 10 seconds apart are within 0.1 per cent of each other, and for plethysmography, a signal corresponding to opening of the mouth shutter ends the test. A summary of the results of each trial is available, and the technician is required to select the results to be entered into the data base and the pulmonary funcition report. A hard copy of the graphics display can be made to manually check the computer calculations; it is saved for future reference.

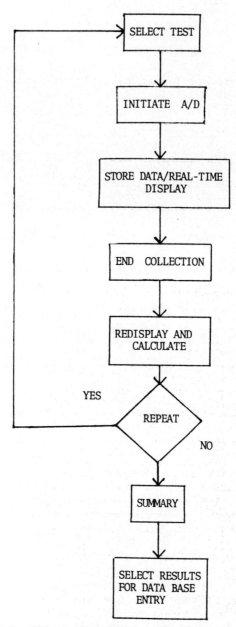

Figure 5.   Flow diagram for the Pulmonary Function
Laboratory Application.   The technician selects the
type of test (spirometry, helium dilution, single breath oxygen, etc.) and initiates data collection with a
simple mnemonic command.   Data are displayed in real
time on his graphics CRT until a criterion (such as a
time limit) is met.   At this point, the CRT is erased
and the data redisplayed along with the calculated variables.

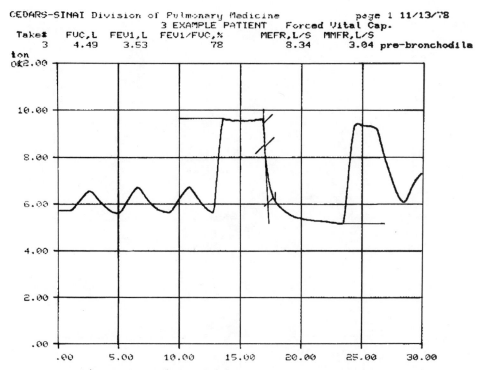

Figure 6. Timed vital capacity maneuver. Two horizontal lines indicate the points from which the forced vital capacity was calculated. A tangent is drawn through the steepest portion of the volume-time tracing and used to define the start of expiration (t0).

The tick marks indicate regions of the spirogram used for the calculation of forced expiratory flows and volumes at (1) 200-1200 ml and (2) 25-75% of the forced vital capacity, and (3) one second after t0. Time is on the horizontal axis (0-30 seconds) and volume is on the vertical axis (0-12 liters).

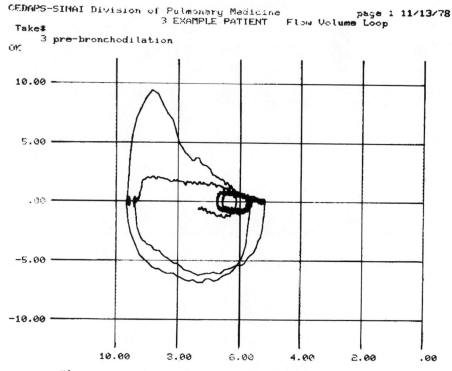

CEDARS-SINAI Division of Pulmonary Medicine   page 1 11/13/78
                    3 EXAMPLE PATIENT   Flow Volume Loop
    Take#
        3 pre-bronchodilation
OK

Figure 7. Flow volume loop obtained from the data
recorded in Figure 6.   Flow is on the vertical axis
(+-10 liters/sec), and volume is on the horizontal axis
(10 to 0 liters).

An exception to the scheme outlined in Figure 5 is the body
plethysmography program. The real-time plot is similar to that obtained by conventional techniques using a constant-volume body
plethysmograph (14,15), but to avoid the problem of baseline drift
in the box pressure signal, the derivatives of the two pressure
signals and the mouth flow signal are stored and used in the calculations of functional residual capacity and airway resistance.

Another difference is in the duration of a trial. For functional residual capacity measurements, the analog signals are sampled, displayed, and discarded prior to receiving a signal indicating that the mouth shutter is closed. For airway resistance measurements, the technician indicates that panting has started by hitting any key on his graphics terminal. After the signal or keystroke indicates the start of a maneuver, the real-time display continues, but only the time derivatives of the signals are calculated and stored on disk. The trial is ended by a signal which corresponds to the opening of the mouth shutter. The results are corrected for body density and printed on the display; lines showing the corresponding slopes are drawn through the loops (plots of box pressure vs. either mouth pressure or mouth flow) (Figure 8). As is the case for the other tests, the technician selects the best trial from a summary for use in the report and entry in the data base.

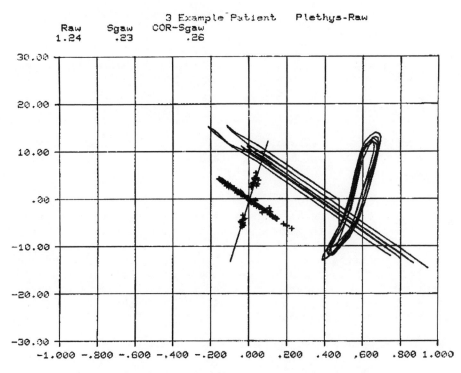

Figure 8. Measurement of airway resistance and specific conductance using a constant volume body plethysmography. The loops are plots of mouth flow and pressure as functions of box pressure, and the crosses represent the derivatives of the mouth flow and pressure signals plotted against the derivative of the box pressure signal. The Y-axis is either mouth pressure (cm H2O) or mouth flow (liters per second X10) and the x-axis is box pressure expressed as an equivalent volume change in, liters.

```
PULMONARY PHYSIOLOGY LABORATORY    Name:    EXAMPLE PATIENT
   Cedars-Sinai Medical Center     Hosp.#: 126311    male        Age: 59
       8700 Beverly Blvd.          Height: 66 in.  Weight: 195 lbs.
     Los Angeles, CA 90048         Tested: 5/07/79  1023 hrs
         (213) 855-6165

     SPIROMETRY    Bronchodilator            Before       After
                        units Pred    Meas %Pred   Meas %Pred  %Change
Slow Vital Capacity       L   3.74    3.10   82    3.48   93     12
Forced Vital Capacity     L   3.74    3.10   82    3.46   92     11
Forced Exp Volume         L   2.90    1.97   67    2.40   82     21
FEV1 / FVC                %    77             63           69
Forced Exp Flow 25-75%   L/s   3.61     .67   18    1.29   35     92
Forced Exp Flow 200-1200 L/s . 5.51   3.85   69    5.56  100     44
Insp Capacity             L   3.03    2.57   84    2.81   92      9
Exp Reserve Vol           L    .71     .53   74     .67   94     26
Maximum Voluntary Vent.  L/M  128             89   125    97     40


     LUNG VOLUMES      units Pred Meas %Pred   PLETHYSMOGRAPHY        Pred Meas
Func Res Cap Ply          L   2.41 3.50  145  Airway Resistance       1.7  2.0
Func Res Cap He           L   2.41 3.38  140  Specific Conductance    .24  .14
Residual Volume           L   1.70 2.97  174
Total Lung Cap            L   5.44 6.07  111
RV / TLC                  %    31   48
     DIFFUSION (CO)                           DISTRIBUTION OF INSPIRED GAS
Diffusing Cap  ml/min/mmHg 27.2 39.1  143  Single Breath O2  <  2.0  1.8
Alveolar Volume        L  5.44 5.70  104  Helium Equil Time  <  2.0  5.2
DL / VA                   5.00 6.85  137
     BLOOD GASES time     pH  PCO2    PO2  HCO3   SO2  FIO2  A-a
                    1015 7.38   43     70    26    94    21   26
```

Figure 9. Pulmonary function report generation from the analog data, including the blood gas data which are automatically appended to the report.

At the conclusion of testing, a preliminary pulmonary function report (Figure 9) is generated and available for inclusion in the patient's chart before he leaves the laboratory. Provisions are made for entering the patient's height, weight, sex, and age so that predicted values can be calculated and included in the report automatically. If an arterial blood gas sample was taken during a predetermined time window, it is automatically included in the report and a link to the ABGS file is stored in the PFRS file.

Exercise Laboratory Application

Eight analog input signals (Table 1) are obtained from the instrumentation in the exercise laboratory, and the variables which can be outputted are listed in Table 2. Up to eight of these parameters can be displayed breath-by-breath in real-time on an 8

TABLE 1 - EXERCISE INPUTS AND SOURCE

1. INSPIRATORY & EXPIRATORY FLOW
   (PNEUMOTACHY GRAPH)

2. RESPIRED CO_2 & O_2 (MASS SPECTROMETER)

3. ARTERIAL BLOOD PRESSURE (TRANSDUCER)

4. O_2 SATURATION (EAR OXIMETER)

5. HEART RATE (CARDIOTACHOMETER)

6. EVENT MARKER

TABLE 2 - EXERCISE OUTPUTS

1. EVENT MARKER (M)

2. RESPIRATORY RATE (f)

3. TIDAL VOLUME (VT)

4. MINUTE VENTILATION ($\dot{V}_E$ ATPS & $\dot{V}_E$ BTPS)

5. CO_2 PRODUCTION ($\dot{V}CO_2$)

6. O_2 UPTAKE ($\dot{V}O_2$)

7. PEAK FLOW (PIF & PEF)

8. END-TIDAL PCO_2 (PET CO_2)

9. END-TIDAL PO_2 (PET O_2)

10. HEART RATE (HR)

11. WORK RATE (WR)

12. MEAN BLOOD PRESSURE (SYSTOLIC, DIASTOLIC, MEAN)

13. O_2 SATURATION (SO_2)

14. INSPIRATORY/EXPIRATORY TIME (T_I/T_E)

15. INSPIRATORY TIME/BREATH DURATION (T_I/T_T)

16. TIDAL VOLUME/INSPIRATORY TIME (V_T/T_I)

17. O_2 UPTAKE/HEART RATE ($\dot{O}_2$ PULSE)

18. VENTILATORY EQUIVALENTS ($\dot{V}EQ$ CO_2 & $\dot{V}EQ$ O_2)

channel analog recorder via the D/A converters (5). Since the input signals are stored on a disk buffer, the outputs can also be displayed at a later time. It is possible to adjust the amplitude scaling and the time scale of the data as they are outputted on the analog recorder.

Two data files are generated for each exercise study. One file contains the raw breath-by-breath data obtained during the study; up to 20 separate studies can be stored on disk. A provision is made for archiving this file onto magnetic tape; the studies can be reloaded back onto disk for subsequent review. The other file (EXERCISE) is a part of the data base and contains a summary of the results of the study, including average values of the calculated output variables for each work rate in the study. The EXERCISE file also contains links to the blood gas file, so that arterial blood gas data can be appended to the report generated at the conclusion of the study.

TABLE 3 - INTENSIVE CARE UNIT INPUTS

1. DATE, TIME	11. VITAL CAPACITY
2. WEIGHT	12. NEGATIVE INSPIRATORY FORCE
3. HEIGHT	13. VD/VT (DEAD SPACE/TIDAL VOLUME)
4. TEMPERATURE	14. CARDIAC OUTPUT
5. RESPIRATORY RATE	15. PULSE RATE
6. TIDAL VOLUME	16. ARTERIAL BLOOD PRESSURE
7. PEAK INSPIRATORY PRESSURE	17. RIGHT ATRIAL PRESSURE
8. STATIC (PLATEAU) PRESSURE	18. PULMONARY ARTERY PRESSURE
9. POSITIVE END-EXPIRATORY PRESSURE	19. PULMONARY CAPILLARY PRESSURE
	20. HEMOGLOBIN
10. VENTILATOR MODE	

TABLE 4 - INTENSIVE CARE UNIT OUTPUTS

1. CARDIAC INDEX	7. pH, PCO_2, PO_2, FIO_2, a/A
2. SYSTEMIC VASCULAR RESISTANCE	8. ARTERIAL OXYGEN CONTENT
3. PULMONARY VASCULAR RESISTANCE	9. MIXED VENOUS OXYGEN CONTENT
4. STROKE VOLUME INDEX	10. OXYGEN DELIVERY
5. LEFT VENTRICULAR STROKE WORK INDEX	11. ARTERIOVENOUS OXYGEN DIFFERENCE
6. RIGHT VENTRICULAR STROKE WORK INDEX	12. COMPLIANCE (EFFECTIVE)

CEDARS – SINAI MEDICAL CENTER

Pulmonary Division Patient Questionaire

Patient
Name   Last   Middle   First

Completed by

SOC. SEC. NO.

Date of completion

Entered by   Date of entry

MEDICAL RECORD NUMBER

[ ]	0	1	2	3	4	5	6	7	8	9
[ ]	0	1	2	3	4	5	6	7	8	9
[ ]	0	1	2	3	4	5	6	7	8	9
[ ]	0	1	2	3	4	5	6	7	8	9
[ ]	0	1	2	3	4	5	6	7	8	9
[ ]	0	1	2	3	4	5	6	7	8	9

Sex;   Male   Female

[ ]	0	1	2	3	4	5	6	7	8	9
[ ]	0	1	2	3	4	5	6	7	8	9
[ ]	0	1	2	3	4	5	6	7	8	9
[ ]	0	1	2	3	4	5	6	7	8	9
[ ]	0	1	2	3	4	5	6	7	8	9
[ ]	0	1	2	3	4	5	6	7	8	9
[ ]	0	1	2	3	4	5	6	7	8	9
[ ]	0	1	2	3	4	5	6	7	8	9
[ ]	0	1	2	3	4	5	6	7	8	9

I. PATIENT PROFILE

ETHNICITY
Black   White   Hispanic
Oriental   Indian   Other

AGE
| 0 | 10 | 20 | 30 | 40 | 50 | 60 | 70 | 80 | 90 |
| 0 | 1 | 2 | 3 | 4 | 5 | 6 | 7 | 8 | 9 |

HEIGHT (inches)
| 0 | 10 | 20 | 30 | 40 | 50 | 60 | 70 | 80 | 90 |
| 0 | 1 | 2 | 3 | 4 | 5 | 6 | 7 | 8 | 9 |

WEIGHT (pounds)
0	1	2	3	4	5	6	7	8	9
0	1	2	3	4	5	6	7	8	9
0	1	2	3	4	5	6	7	8	9

SITE VISIT
RICU   MICU   In-patient
6SE   Clinic   PFT
          Office   Out-PFT

DATE OF ADMISSION OR VISIT
Mo.   1   2   3   4   5   6   7   8   9   10   11   12
Day   0   10   20   30
           0   1   2   3   4   5   6   7   8   9
Year   78   79   80   81   82   83   84   85   86   87

DATE OF:
Mo.   1   2   3   4   5   6   7   8   9   10   11   12
Discharge   Day   0   10   20   30
Death   0   1   2   3   4   5   6   7   8   9
Autopsy   Y   N   78   79   80   81   82   83   84   85   86   87

II. HISTORY

COMPLAINTS:
If "None" marked, skip to PPD status
None   Dyspnea   Cough
Chest pain   Hemoptysis   Wheezing
Abnormal CXR   Expectoration   Fever
Orthopnea   PND   Other

DYSPNEA   (Quality)   Min   Mod   Severe   Very severe   Oth
          (Duration)   < 1 wk   1 wk - 3 mos   3-12 mos   > 1 yr

COUGH   (Frequency)   Occasional   Frequent
        (Duration)   < 1 wk   1 wk - 3 mos   3 mos x 2 yrs or 6 mos/yr

CHEST PAIN   Pleuritic   Angina   Musculoskeletal   Non-spec

HEMOPTYSIS   (Quantity)   Questionable   Definite
                    > 30 cc   > 600 cc
             (Duration)   < 1 wk   > 1 wk   Mult. episodes

WHEEZING   Rarely   With resp. infect. only
           Frequent   Post - exercise

EXPECTORATION   (Quantity)   ≤ 1 tsp   2 tsp - ½ cup   > ½ cup daily
                (Quality)   Clear   Purulent (yellow-green)
                (Duration)   < 1 wk   1 wk - 3 mos   3 mos x 2 yrs or 6 mos/yr

FEVER   (Duration)   < 1 wk   1 wk - 3 wks   > 3 wks.

PPD STATUS   Neg   Pos   Unknown

FAMILY HISTORY
Neg   Allergy   COPD   TBC
Carcinoma   Interstitial disease   Asthma
Cardiac or cardiovascular   Renal   Other

PAST RESPIRATORY ILLNESS
None   TBC   Pneumonia
COPD   Asthma   Sarcoidosis
Carcinoma   Fibrosis   Pulm emboli
Thoracic surgery   Other

(Right margin line numbers: 1–64)

Figure 10.   Page one of four pages of the patient questionnaire.

Questionnaire Application

Each patient seen by the pulmonary division has a questionnaire (Figure 10) completed, which is entered into the computer using either an optical mark sense form or manually from a CRT. The form is four pages long and contains over 500 separate items, including information on history, physical examination, laboratory data, chest radiograph, biopsy procedures and final diagnoses. The latter may be selected from a list comprising 230 diagnoses. The questionnaire data reside permanently on disk and multiple questionnaires can be answered for any patient, so that changes in his medical status can be recorded.

Intensive Care Unit Application

Ventilatory and hemodynamic intensive care unit data entered into the data base file ICUS. As the data are entered, the arterial blood gas file is automatically searched for blood gas results which fall within a predetermined time window of the time data were taken; when blood gas data that meets this criteria is found, a link is stored in the ICUS file. This process results in the generation of a report which integrates the blood gas with the hand entered RICU data. Table 3 lists the input data and Table 4 the calculated output data. Both input and output data can be displayed in tabular form for each patient.

Database Searches

In general, the file structure is designed to facilitate the retrieval of individual patient data by patient number (i.e. blood gas reports, pulmonary function reports, questionnaires, etc.). Searches by other criteria may require more extensive disk activity and are somewhat more time consuming. Those routines which are used to generate standard reports are implemented by simple mnemonic commands to the appropriate programs are loaded into the user's terminal partition.

Other types of searches which are not done on a routine basis are easily implemented using various search mechanisms which are incorporated into the software. These include the capability to search a file(s) for values under the following conditions:

(value) WITHIN or NOT WITHIN

(value or date) =, < or >; NOT =, NOT < or NOT >

(date interval) WITHIN or NOT WITHIN

These conditions may be strung together to either select or reject file records based on a series of criteria. The works SELECT and REJECT determine wether the record is saved or discarded by the search routine. Each record in the file is tested by the search conditions in sequence; a "hit" on a SELECT condition causes the record to be saved and the remainder of the conditions ignored. For example the sequence:

(field name 1) (value 1)< REJECT (field name 2) (value 2)> SELECT

will search the file and save all records for which (field name 1) is greater than (value 1) AND (field name 2) is greater than (value 2), making this an implicit AND search (must meet both criteria). In this sequence, replacing REJECT with SELECT would create an implicit OR search, since the search would save all records with either (field name 1)<(value 1) or (field name 2)>(value 2).

The use of these mechanisms requires a superficial knowledge of FORTH. They allow the user to quickly initiate many different types of searches, and provide outputs in the form of tables, histograms, or plots on graphics terminals. Simple statistics (mean, standart deviation, t-test, chi-square) are also available to facilitate data analysis.

FOOTNOTES

1. The software was developed to our design specifications by FORTH, Inc. 815 Manhattan Ave., Manhattan Beach, Ca. 90266. Other sources for the language are DECUS (Digital Equipment Users Society, 126 Parker St., Maynard, Ma. 01754) and the FORTH Interest Group, 787 Old Country Road, San Carlos, Ca. 94070.

REFERENCES

1. NAIMARK, A., CHERNIACK, R.M., PROTTI, D.: Comprehensive respiratory information system for clinical investigation of respiratory disease. American Rev Res Disease 103:229-239, 1971.

2. PACK, A.I., MC CUSKER, R., MORAN, F.: A computer system for processing data from routine pulmonary function tests. Thorax 32(3):333-341, 1977.

3. SOTO, R.J., FORSTER, H.V., RASMUSSEN, B.: Computerized method for analyzing maximum and partial expiratory flow-volume curves. J. Appl. Physiol. 39(2):315-317, 1975.

4. HOFFER, E.P., KANAREK, D., KAZEMI, H.: Computer interpretation
    of ventilatory studies. Comp. and Biomed. Res.,
    6:347-354, 1973.

5. BEAVER, W.L., WASSERMAN, K., WHIPP, B.J.: On-line computer an-
    alysis and breath-by-breath graphical display of exercise
    function tests. J. Appl. Physiol. 34:128-132, 1973.

6. HERR, G.P.,SULLIVAN, S.F., KHAMBATTA, H.J.: On-line acquisi-
    tion of pulmonary gas exchange data. IEEE Transactions
    on Biomed. Engr., BME-25(1): 1978.

7. CARDUS, D. NEWTON, L.: Development of a computer technique
    for the on-line processing of respiratory variables.
    Comput. Biol. Med. 1:125-133, 1970.

8. OSBORN, J.J., BEAUMONT, J.O., RAISON, J.C., AND GERBODE, F.:
    Measurement and monitoring of acutely ill patients by di-
    gital computer. Surgery 64:1058-1070, 1968.

9. LEAVITT, M.B., LEINBACH, R.C.: A generalized system for colla-
    borative on-line data collection. Computers and Biomed.
    Res. 10(4):413-421, 1977.

10. COLLEN, M.F.: General requirements for a medical information
    system (MIS). Computers and Biomed. Res. 3:393-406,
    1970.

11. HICKS, S.M.: FORTH's forte is tighter programming.
    Electronics 52(6): 114-118, 1979.

12. RATHER, E.D. AND MOORE, C.H.: The FORTH approach to operat-
    ing systems. ACM '76 Proceedings. 233-240:1976.

13. The timing of the forced vital capacity. Amer. Rev. Resp.
    Dis., 119:315-318, 1979.

14 DUBOIS, A.B., BOTELHO, S.Y., COMROE, J.H.: A new method for
    measuring airway resistance in man using a body plethys-
    mograph: Values in normal subjects and in patients with
    respiratory disease. J. Clin. Invest. 35:327-335,
    1956.

15. DUBOIS, A.B., BOTELHO, S.Y., BEDELL, G.N.: A rapid plethysmo-
    graphic method for measuring thoracic gas volume: A com-
    parison with a nitrogen washout method for measuring
    functional residual capacity in normal subjects. J.
    Clin. Invest. 35:322-326, 1956.

A NEW APPROACH TO APPLICATION OF A COMPUTER IN SPIROMETRY

A. Patel, Ph.D*, D. Tashkin, M.D.**,
G. Ellis, B.A.* and E. Peterson, B.S.*

*Cavitron Corporation
1350 Ave. of the Americas
New York, New York 10019

**UCLA School of Medicine
Los Angeles, California 90024

ABSTRACT

The traditional approach of use of computers in pulmonary diagnostic medicine in general and spirometry in particular has been
to interface an existing spirometer to a general purpose computer.
The problems and limitations of this traditional approach are discussed. The advantages of the computer thus have not been readily
available outside a few university-affiliated hospitals. Remaining
large segments of the medical community still continued spirometry
with manual data reduction procedures which are both time-consuming
and prone to errors. These manual data reduction errors are also
briefly discussed.

Recently the professional community is increasingly recognizing the need for both greater accuracy in spirometry and for its
more widespread use for early identification of patients with
chronic respiratory diseases. Our new approach to computerization,
while eliminating manual procedures, avoids the complexity, bulk,
etc., pitfalls of the traditional approach. This approach leads to
quick and easy as well as accurate computerized spirometry.

In our approach a pre-programmed micro-computer is integrated
into a self-contained instrument. Nine push-button keys provide
all necessary user controls. Each key has a single well-defined
function which is labeled by a meaningful English word. Each key
stroke is confirmed by a clear alpha-numeric display in the common

language of spirometry. The display also shows the results of spirometric tests. The instrument permits testing of FVC and MVV maneuvers. Flow-volume loop capability is included in the FVC manuever. Widely-used test parameters and predicted normal values are computed immediately and displayed on demand. The instrument meets critical requirements of the currently proposed spirometry standards. The back-extrapolation method is used to determine the start of timed volumes (FEVt). The end point criterion for termination of FVC measurement is more conservative than currently proposed standards. Breathing rate is also computed for the MVV results. Corrections due to ambient temperature, barometric pressure and relative humidity are easily provided. The calibration as well as its check is easily achieved with a hand syringe.

A bi-directional pneumotach is the flow sensor. A new chopper stabilized principle permits use of a single flow sensor for the entire dynamic range. A voltage-to-frequency converter is used for analog to digital conversion. A micro-processor, memory and LED display are integrated on a single circuit board. Both analog and serial digital outputs are provided for printed and/or graphic records. The computerized spirometer weighs only 15 lbs. and can thus also be easily used at bedside.

A comparative evaluation with traditional volumetric spirometers requiring manual procedures is presented. Test data includes linearity, accuracy and frequency response measurements for volume and flow parameters.

INTRODUCTION

The traditional approach of the use of computers in pulmonary diagnostic medicine in general and spirometry (1-3) in particular has been to interface an existing spirometer to a general purpose computer. This required the clinician, who is usually a pulmonary physician or technologist, to solve problems of interfacing and programming of the computer. His only alternative was to seek the services of a specialist such as a bio-medical, computer or electrical engineer with the required skills. Even having solved these problems, the user was required to communicate with the computer through a keyboard or console which gave the impression of great complexity, and thus discouraged most clinicians. Also such an approach led to bulky, non-mobile and engineer dependent expensive equipment. The consequence is that such activity is restricted to a few university-affiliated or research-oriented medical centers. Remaining large segments of the medical community still continue spirometry with manual data reduction procedures as applied to a spirogram.

Manual data reduction procedure(4) requires finding a point on the curve with the steepest slope and drawing this slope in addition to several interpolations for volume and time measurements. Subsequently, many subtractions, multiplications and divisions are required to obtain commonly-used volume and flow parameters of spirometry. Such activity is not only time-consuming, but prone to errors. It has been shown that manual data reduction of the same spirogram by different clinicians or even by the same clinician at different times yields variability.(5-6) This variability in spirometry, unlike that due to patient effort(7-8) or spirometer design(9-11), can be solved with automation. Early attempts to automate spirometry with electronic spirometers(12) could not achieve the necessary(13) backward extrapolation method of starting point and strict criteria of end point. The computer is uniquely suitable to solving these problems and makes possible spirometry with a speedy, repeatable and accurate method of computation.

Recently the professional community has increasingly recognized the need for both greater accuracy(13) in spirometry and for its more widespread use(14) for early identification of patients with chronic respiratory diseases. This paper presents a new approach to application of a computer in spirometry which avoids the complexity, bulk, etc. pitfalls of the traditional approach, while eliminating manual procedures. This approach leads to quick and easy, as well as accurate computerized spirometry.

OBJECTIVES

Our approach is defined by the following objectives.

1. Evolve an integrated self-contained microcomputer-based spirometer as small as possible which would meet the requirements(13) for Forced Vital Capacity(FVC) and Maximal Voluntary Ventilation (MVV) tests. FVC test should permit traditional volume-time spirogram, as well as complete flow-volume loop maneuvers.

2. Minimize the number of push-button keys or controls necessary for the operator.

3. Simplify the procedure of operation by use of only the English language or the well-known abbreviations of spirometry such that any pulmonary physician or technologist can feel at ease with the computerized spirometer.

MATERIALS AND METHODS

Since transducers for flow-measuring are much smaller compared to those for measuring volume, we selected a new, improved flow transducer(15) for spirometry. The instrumentation for spirometry

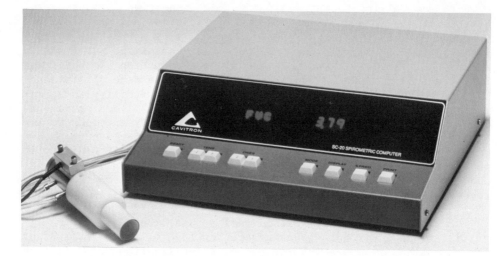

Fig. 1 A Spirometric Computer (SC-20). The instrumentation for a flow sensor has been integrated
with a microcomputer to achieve this compact (14-1/2"W
x 14"D x 4-1/2"H, 15 lb.) device.

was integrated with an Intel 8080 microprocessor-based microcomputer to form a small (14-1/2" W x 14" D x 4-1/2" H, 15 lb.) spirometric computer as shown in Fig. 1. The software and hardware details of the spirometric computer (SC) are given below, referring
to the device as SC-20 for the rest of the text.

Keys and Controls

A power on-off switch, three potentiometric calibration controls and connectors for data-recording are provided on the back
panel of the SC-20. The front panel as shown in Fig. 1 forms the
operator-interface which consists of nine push-button keys and a
clearly visible alphanumeric display.

Each key has a single, well-defined function which is labelled
by a meaningful English word. Even though there are nine keys,
functionally there are only six groups of keys since the four keys
for entering numerical data are utilized as a group only during patient data entry and periodic calibration.

Going from left to right in Fig. 1, the READY key, when depressed, activates the SC-20 when any of the testing modes have
been selected. Subsequent patient exhalation in the flowhead will
initiate gathering of flow and volume data. The next four keys,
TENS down, TENS up, ONES down and ONES up, form a data entry group

of keys. They economically permit entering patient or ambient environmental data by increasing or decreasing the current data value
by ten or by one. The MODE key halts any activity in progress and
causes the display to indicate current mode. A subsequent mode key
stroke causes the instrument to sequentially select the next mode.
The sequence of five modes consists of Patient Data (PTD), Forced
Vital Capacity (FVC), Maximal Voluntary Ventilation (MVV), Plot
(PLT) and Calibration (CAL). The DISPLAY key sequentially selects,
for display the various parameters which have been measured or calculated or required for entry for a specific mode. The % PRED.
key displays the current test parameter as a percentage of its appropriate predicted normal value which is used from the memory of
the microcomputer. If no predicted normal value is available, the
display is blanked. The last extreme right key is the PRINT key
which activates an externally-connected optional graphic printer
(recorder) for a permanent record. Either volume-time or
flow-volume graph can be plotted by the recorder which also prints
measured, computed and % of predicted data.

Software

Each key stroke is confirmed by the alphanumeric display in
simple abbreviations of English language or standardized spirometric terms. This visual key stroke confirmation design, along with
a predetermined sequential stepping scheme of the software, as
shown in Fig. 2, yields a simplified procedure of operation for any
clinician. The stroked key or the event of the patient maneuver
leading to each of the preprogrammed steps represented by a block
is identified in Fig. 2. Each bloc (step) shows the resultant alphanumeric display in the top half and its explanation in the bottom half. It is obvious in Fig. 2 that MODE, READY and DISPLAY
keys are prominent in the sequential stepping scheme of the
software.

Clinician-Computer Interaction

In order to appreciate the simplicity of the operation of this
clinician-computer interface, the details of sequential steps for
FVC test on a patient are given below:

The MODE key stroke which sequentially steps the SC-20 through
the five modes as shown on the top row of Fig. 2, is used to select
the desired mode. It is assumed here that the SC-20 was earlier
calibrated for bi-directional use in the calibration mode with a
four-liter hand syringe and appropriate entries for room temperature, barometric pressure and humidity. Before performing the FVC
test for a patient, first the MODE key is used to select the patient data entry (PTD) mode. Then in this mode the DISPLAY key is

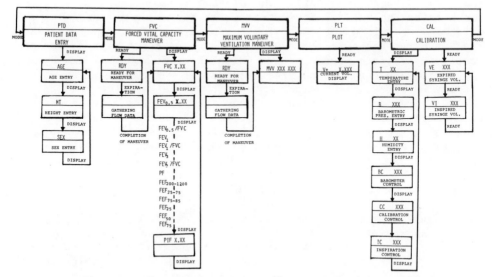

Fig. 2   The software block diagram of the SC-20.
This diagram represents a predetermined sequential
stepping scheme which provides a simplified procedure
of operation as described in the text.

sequentially stepped (see Fig. 2) and patient°s age, height and sex
entries are respectively entered by the operator as they are asked
by the display.   The data entry group of keys is used for these entries.   In the sex entry step the initial male (M) sex display can
be altered to female (F) or further altered back to male and so   on
by depressing   any one of the four data entry groups of keys.   The
patient data entry mode steps can be usually carried   out   quickly.
Approximately a minute thus spent is required for use of appropriate predictive normals for subsequent computations of data   during
the spirometric modes.

CAVITRON SC-20 SPIROMETRIC COMPUTER
FLOW-VOLUME LOOP

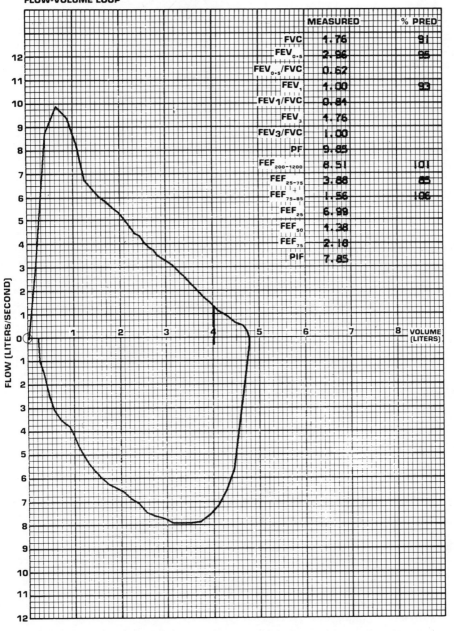

MEASURED		% PRED
FVC	4.76	91
FEV$_{0.5}$	2.96	95
FEV$_{0.5}$/FVC	0.62	
FEV$_1$	4.00	93
FEV$_1$/FVC	0.84	
FEV$_3$	4.76	
FEV$_3$/FVC	1.00	
PF	9.85	
FEF$_{200-1200}$	8.51	101
FEF$_{25-75}$	3.88	85
FEF$_{75-85}$	1.56	106
FEF$_{25}$	6.99	
FEF$_{50}$	4.38	
FEF$_{75}$	2.18	
PIF	7.85	

Fig. 3   The flow-volume graph and   the numerical
results of the spirometric computations both being recorded by a graphic printer.

The MODE key is used again to sequentially step to the next FVC mode. In the FVC mode, the READY key is pressed which prepares the SC-20 for gathering flow and volume data. Then the patient with nose clips exhales through the mouth-piece of the flow sensor with maximal effort as per the instructions and training given to him. The expiration by the patient initiates the SC-20 for collection of data and at the completion of the maneuver, the FVC value appears in the display. Other test measurements and computations can be displayed in the sequential stepping manner shown in Fig. 2.

If an optional graphic printer is used, then the PRINT key stroke activates it. Both graphic and numerical data is reproduced for hard copy as shown in Fig. 3 from the 1K Byte of read-write memory of the SC-20. Such data retrieval from the memory overcomes the slewing rate limitation of the traditional graphic X-Y plotters.

The entire FVC test for the patient thus is rapidly accomplished and the automation provides immediately all the flow and volume parameters printed in Fig. 3. For the computations of these results, the microcomputer of the SC-20 not only executes the back extrapolation method of starting point for FEVt, but also has a strict end of test criteria which prevents premature termination of the volume-time maneuver for FVC test.

The operation of MVV, plot and calibration modes can now be easily grasped in Fig. 2. For the MVV test, MVV and breathing rate are computed. The plot mode provides direct flow and volume analog outputs without interpretation by the microcomputer.

HARDWARE

The hardware block diagram of the SC-20 is shown in Fig. 4. All of the digital electronics blocks of the microcomputer, including the display and the keyboard interface, are integrated onto a single printed circuit board. An Intel 8080 microprocessor is used with the program for it being stored in 6K Bytes of read-only memory (ROM). A scratch pad random access memory (RAM) of 1K Byte is used for data collection, computation and retrieval. This size of RAM memory is adequate because the software uses a selective data compression technique during the data processing. A voltage-to-frequency converter is used for analog-to-digital conversion. Real-time processing of the flow data is accomplished by the use of vectored interrupts. The interrupt signal to the mi-

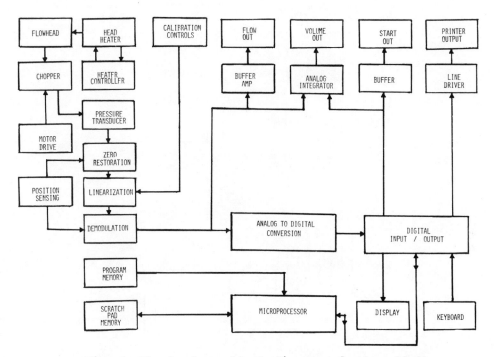

Fig. 4 The hardware block diagram of the SC-20.
This diagram represents the hardware for microcomputer
and components for spirometry by a new flow sensor.

crocomputer is at each 1.25 milliliter of volume measured by the
flow sensor. An electronic 40 Hz clock provides pulse at every 25
millisecond. Thus 1.25 milliliter and 25 milli-second are the primary units of measurements in the microcomputer. All measured data
and all derived parameters are represented as exact integers in
these primary units. Only the final results are converted to
liters, seconds and minutes, thus avoiding arithmetic round-off
problems.

In Fig. 4, in addition to hardware for the microcomputer, the
components of a new chopper stabilized bi-directional pneumotach

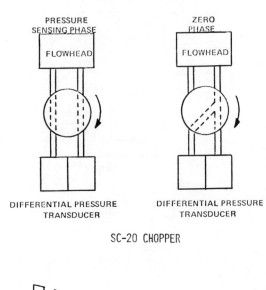

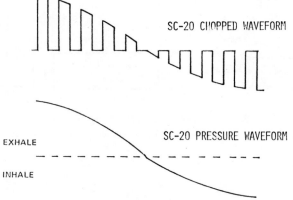

Fig. 5 A new stabilization principle for the bi-directional pneumotach of the SC-20. A rotating pneumatic switch called "chopper" is represented by the circle and the arrow. The pneumatic switching between the pressure sensing phase and the zero phase provides the cancellation of the drift of the pressure transducer.

flow sensor are also represented. In the traditional pneumotach, the flowhead is directly connected to a precision differential pressure transducer. Such a method is known to have a drift problem for measurement of low pressures corresponding to low flows. Since pneumatic resistance of the flowhead cannot be arbitrarily increased, such a drift problem results into the inability of the traditional pneumotach to measure both high and low flows accurately. Fig. 5 illustrates the chopper stabilization principle used in the pneumotach of the SC-20 which overcomes the above-stated limitation of the traditional pneumotach. A rotating pneumatic switch called a "chopper" is inserted between the patient flow head and the differential pressure transducer. The SC-20 chopper rotates at a rate of 30 cycles per second, alternating between the pressure sensing phase and the zero phase as shown in the upper section of Fig. 5. During the zero phase of the chopper the sensing tubes of the differential pressure transducer are connected together to the outlet port and thus provide an effective zero flow reference point. Any drift of the differential pressure transducer is common during both phases of the chopper and is thus automatically cancelled out. The bottom of Fig. 5 illustrates the chopped waveform and the corresponding continous pressure waveform. The chopper stabilized pressure waveform serves as the input to the microcomputer via the analog-to-digital conversion.

The above described software and hardware features thus provide a simplified microcomputer-based spirometer.

TABLE 1

ASSESSMENT OF LINEARITY - VOLUME MEASUREMENT

TEST #	VOLUME MEASUREMENT IN LITERS		ERROR IN	
	STEAD-WELLS	SC-20	mL	%
1	1	0.98	-20	-2.00
2	2	1.99	-10	-0.50
3	3	2.98	-20	-0.67
4	4	3.96	-40	-1.00
5	5	4.93	-70	-1.40
6	6	5.93	-70	-1.17
7	7	6.90	-100	-1.43
8	8	7.93	-70	-0.88
			AVG. -50 mL	-1.13%

ASSESSMENT OF LINEARITY - VOLUME MEASUREMENT

SCHEME:

STEAD-WELLS   SC-20

THE TWO SPIROMETERS WERE CONNECTED IN-SERIES WITH
STEAD-WELLS ACTING AS A VARIABLE VOLUME GENERATOR.

ASSESSMENT OF LINEARITY - FLOW MEASUREMENT

SCHEME:

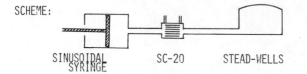

SINUSOIDAL   SC-20   STEAD-WELLS
  SYRINGE

THE SINUSOIDAL SYRINGE GENERATOR OF KNOWN VOLUME AND
VARIABLE FREQUENCY SUPPLIED SEVERAL DYNAMIC FLOW RANGES FOR THE
SERIES CONNECTED SPIROMETERS. AT A SELECTED FREQUENCY, THE FLOW
RANGED BETWEEN ZERO AND KNOWN PEAK FLOW. THE LINEARITY OF THE
FLOW MEASUREMENT BY THE SC-20 SPIROMETER IN EACH DYNAMIC FLOW
RANGE AND THE ACCURACY OF THE SUBSEQUENT INTEGRATION WAS ASSESSED
BY THE ACCURACY OF THE MEASURED VOLUME OF THE SYRINGE.

Fig. 6 The schemes for comparative assessment of the
linearity of volume and flow measurements by a SC-20
and a Stead-Wells spirometer.

RESULTS

The SC-20 meets volume-and-flow ranges and other requirements
of the proposed standards.(13) Validation of a spirometer by bench
tests requires the assessments of the linearity of volume and flow
measurements, the dynamic frequency response, the starting point
for FEVt, and the end of test criteria for FVC spirogram.

Linearity of volume and flow measurements:

A Stead-Wells spirometer at room temperature acted as a variable volume generator as shown in the scheme of Fig. 6 for assessment of linearity of volume measurements of a series connected
SC-20 spirometer. The Stead-Wells spirometer consists of a
ten-liter non-counterweighted water-sealed plastic bell with an attached pen which directly records on a rotating kymograph.

TABLE 2

ASSESSMENT OF LINEARITY – FLOW MEASUREMENT

TEST #	PEAK FLOW MEASUREMENT IN LITERS/SEC.			FVC MEASUREMENT IN LITERS		
	SYRINGE MODEL	STEAD-WELLS	SC-20	SYRINGE MODEL	STEAD-WELLS	SC-20
1	1.47	1.44	1.45	3.0	2.94	3.00
2	2.21	2.11	2.13	3.0	2.98	2.93
3	3.20	3.15	3.13	3.0	2.99	2.99
4	5.06	4.97	4.95	3.0	3.00	2.99
5	6.22	6.07	6.10	3.0	2.97	3.00
6	7.78	7.60	7.65	3.0	2.97	3.00
7	9.51	9.40	9.35	3.0	2.98	3.00
8	11.02	10.76	10.80	3.0	2.98	3.01
9	12.23	12.00	11.90	3.0	2.99	3.02

	ERROR ANALYSIS OF THE ABOVE DATA	MAXIMUM	AVERAGE
ERROR IN PEAK FLOW MEASUREMENT IN L/S. IF < 0.2 L/S	SC-20 VS STEAD-WELLS	0.10L/S	0.058L/S
	SC-20 VS MODEL	2.69%	0.138L/S
ERROR IN FVC MEASUREMENT IN L IF < 50 L	SC-20 VS STEAD-WELLS	2.04%	22.22 mL
	SC-20 VS MODEL	20mL	8.9 mL

The bell of the Stead-Wells spriometer from a known volume position was dropped, thus serving as a generator of known volume for the series-connected SC-20. The comparative volume measurements given in Table 1 show that the SC-20 has adequate linearity in the examined range up to eight liters.

The scheme for assessment of linearity of flow measurements is also shown and described in Fig. 6. This scheme used a three-liter sinusoidal syringe at room temperature as a generator for the series-connected SC-20 and Stead-Wells spirometers. At a selected frequency for the sinusoidal generator, the flow ranged between zero and known peak flow. The linearity of the flow measurement by the SC-20 spirometer in each dynamic flow range and the accuracy of the subsequent integration was assessed by the accuracy of the measured volume of the syringe. The peak flow and the FVC measurements by the series-connected SC-20 and Stead-Wells spirometers are given in Table 2. Comparative error analysis of these measurements and the syringe model is also given in Table 2. This analysis shows that the SC-20 has adequate linearity of flow measurement in the dynamic range of 0 to 12 liters/second.

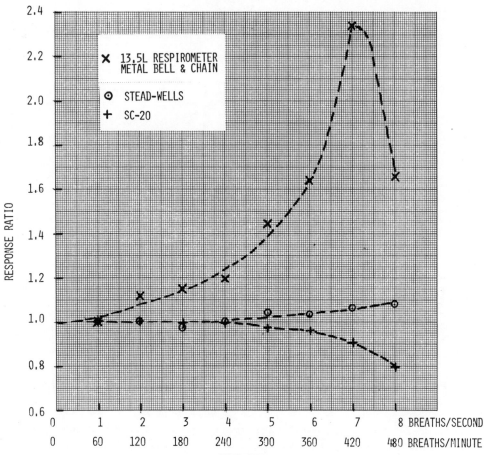

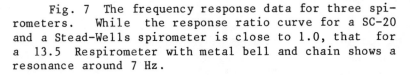

Fig. 7 The frequency response data for three spirometers. While the response ratio curve for a SC-20
and a Stead-Wells spirometer is close to 1.0, that for
a 13.5 Respirometer with metal bell and chain shows a
resonance around 7 Hz.

Frequency Response:

      The frequency response of the SC-20 spirometer was assessed by
using a 140 milliliter sinusoidal generator at room temperature.
This sinusoidal generator provided frequencies of up to 8Hz (i.e.
480 breaths/minute) with the flow well within 0 to 12 liters/second
range.   The MVV mode of the SC-20 was used for the volume measurements at frequencies marked in Fig. 7.   The measured MVV volume was
converted to the measured tidal volume by dividing MVV by the fre-

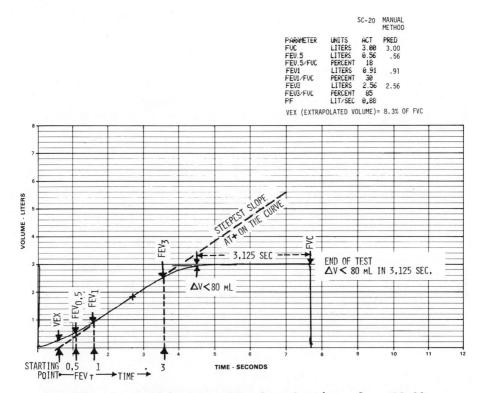

SC-20 MANUAL
METHOD

PARAMETER	UNITS	ACT	PRED
FVC	LITERS	3.00	3.00
FEV.5	LITERS	0.56	.56
FEV.5/FVC	PERCENT	18	
FEV1	LITERS	0.91	.91
FEV1/FVC	PERCENT	30	
FEV3	LITERS	2.56	2.56
FEV3/FVC	PERCENT	85	
PF	LIT/SEC	0.88	

VEX (EXTRAPOLATED VOLUME)= 8.3% OF FVC

Fig. 8 Graphical record of evaluation of a SC-20
for (1) Back extrapolation method of starting point for
FEVt (2). End of test criteria which requires that the
change of volume has to be less than 80 milliliters in
a 3.125-seconds time interval.

quency of breaths per minute. Subsequently, the normalized response ratio shown in Fig. 7 was obtained by dividing all tidal volume measurements by the tidal volume corresponding to frequency of
1 Hz (1 breath/second).

Fig. 7 also shows similarly obtained frequency response of a
13.5 liter respirometer with an indirect metal bell and chain recording system9 which was observed to be resonating around 7 Hz.
The dramatic improvement of the frequency response by the direct
recording plastic bell design(10) of the Stead- Wells is also confirmed in Fig. 7.

Fig. 7 shows that the frequency response of the SC-20 spirometer is within 10% up to 7 Hz and thus adequate according to recently proposed requirements.(13)

Starting point for FEVt and end of test criteria for FVC Spirogram:

The microcomputer of the SC-20 is programmed to use the back extrapolation method(13) for determining the starting point for FEVt measurements.

In order to prevent premature termination of the FVC measurement in a patient with obstructive lung disease, it was recently proposed(13) that the end of test should occur when the volume change in a 0.5 second interval is less than 25 milliliters or the average flow over a 0.5 second interval is less than 50 milliliters per second. The SC-20 utilizes a more strict criteria which requires that in a 3.125 seconds time interval, the change of volume has to be less than 80 milliliters for end of test to occur.

Both the starting point for FEVt and end of test criteria for FVC of the SC-20 was assessed by measurement of a complete stroke of a sinusoidal syringe as shown in Fig. 8. The back extrapolation method of starting point is manually applied to the analog recorded output. The comparative table in Fig. 8 clearly shows that the FEV0.5, FEV1 and FEV3 determinations by manual method agree with those computed by the microcomputer of the SC-20. The end of test criteria of the SC-20 is also confirmed in Fig. 8 by the examination of the analog recorded output of the volume-time curve. The volume change in the final 3.125 seconds was less than 80 milliliters and thus the end of test occurred and the spirogram was terminated as per the design of the SC-20.

CONCLUSIONS

This paper has presented a new approach to the application of a computer which automates spirmetry while meeting the requirements of currently proposed standards. This approach, which simplifies the clinician-computer interaction by the use of a preprogrammed microcomputer, merits consideration for the automation of other pulmonary function laboratory tests.

REFERENCES

1. ROSNER, SW, PALMER, A, WARD, SA, ABRAHAM, S, AND CACERES, CA: Clinical spirometry using computer techniques. Amer. Rev. Resp. Dis., 94: p. 181, 1966.

2. DICKMAN, ML, SCHMIDT, CD, GARDNER, RM, MARSHALL, HM, DAY, WC, AND WARNER HR: On-line computerized spirometry in 738 normal adults. Amer. Rev. Resp. Dis., 100: p. 780, 1969.

3. TSAI, MJ AND PIMMEL, RL: Interactive on-line computerized computation of spirometric parameters (abstract). Proc. 30th Annu. Conf. Eng. Med. Biol., 19: p. 374, 1977.

4. KANNER, RE AND MORRIS, AH: Clinical Pulmonary Function Testing, Intermountain Thoracic Society, Salt Lake City, Utah: pp. 1-7, 1975.

5. ROSNER, SW, ABRAHAM, S, AND CACERES, CA: Observer variation in spirometry. Dis. of the Chest, 48: p. 265, 1965.

6. SNIDER, GL, RIEGER, BA, DEMAS, T AND DOCTOR L: Variations in the measurement of spirograms. Am. J Med. Sci., 254: p. 269, 1967.

7. HANKINSON, JL, AND PETERSON, MR: Data analysis for spirometry instrumentation standards (abstract). Am. Rev. Respir. Dis., 115 (Supplement): p. 116, 1977.

8. GLINDMEYER, HW: An Experimental Investigation of the Errors of Spirometry, Ph.D. Dissertation, Tulane University, New Orleans: p. 187, 1976.

9. STEAD, WW, WELLS, HS, GAULT, NL AND OGANANOVICH, J: Inaccuracy of the conventional water-filled spirometer for recording rapid breathing. J. Appl. Physiol., 14: p. 448, 1959.

10. WELLS, HS, STEAD, WW, ROSSING, TD AND OGANANOVICH, J: Accuracy of an improved spirometer for recording of fast breathing. J. Appl. Physiol., 14: p. 451, 1959.

11. GLINDMEYER, HW, ANDERSON ST, DIEM, JE AND WEIL, H: A comparison of the Jones and Stead-Wells spirometers. Chest, 73: p. 596, 1978.

12. FITZGERALD, MX, SMITH AA AND GAENSLER, EA: Evaluation of "electronic" spirometers. N. Engl. J. Med., 289: p. 1283, 1973.

13. The American Thoracic Society (ATS) Statement–Snowbird Work-
     shop on Standardization of Spirometry. Amer. Rev.
     Resp. Dis., 119:p. 831, 1979.

14. The American College of Chest Physicians (ACCP) Statement –
     Office Spirometry in Clinical Practice. Chest, 74: p.
     298, 1978.

15. DEMPSTER, PT AND PUN, JY: Spirometer apparatus and method.
     United States Patent 3,924,612, 1975.

A NEW CONCEPT OF THE FORCED EXPIRATORY SPIROGRAM AND THE EFFORT

INDEPENDENT PORTION OF THE FLOW-VOLUME LOOP. A COMPUTER ANALYSIS.

George R. Meneely, M.D., Francis S. Knox, III, Ph.D.,
Steven A. Conrad, M.D., Chester Ellis, M.S.,
Stephen G. Jenkinson, M.D., Ronald B. George, M.D.

The Department of Physiology and Biophysics
and the Department of Medicine,
Louisiana State University Medical Center, Shreveport
Shreveport Veterans Administration Medical Center
Shreveport, Louisiana

Supported in part by a grant in aid of research from E.
Rhodes Carpenter and Naples Foundation and from the Francis S.
North Foundation

The forced expiratory spirogram continues to interest those
concerned with pulmonary function as a prctical, fast, reproducible
and (1-6) clinically useful test of ventilatory function.   It has
"withstood the test of time". This interest is not new. While Sir
Humphrey Davy did address the matter of volume of the lung and its
subdivisions in the last decade of the 18th century (7, 8), it was
Hutchinson in 1846 (9) who "...fathered the concept of vital capacity".   Bartels, Bucherl, Hertz, Rodewold and Schwab (10) state
that, so far as known, Volhard (11) in 1908 was the first to observe that, on forced expiration of the vital capacity into a spirometer by a patient with emphysema, the bell rises more slowly
than with a normal person. Raither (12), a pupil of Volhard, made
the first detailed study of the form of the forced expiratory spirogram. (Fig. 1) Tiffeneau and Pinelli (13, 14) introduced the concept of the one second forced expiratory volume, taking it to be
the "pulmonary capacity utilizable with effort".   Gaensler (15) introduced the concept of timed forced expiratory volumes in America,
devised an inexpensive and effective device (16) for measuring
these volumes, introduced the term "timed vital capacity" and drew

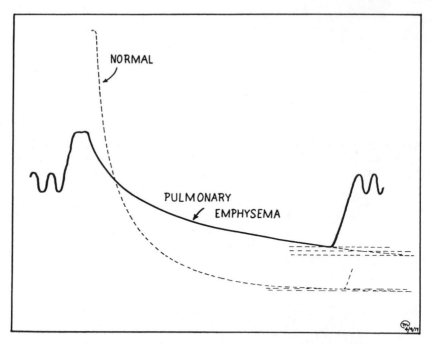

NORMAL

PULMONARY
EMPHYSEMA

FIGURE 1.   A spirogram from a normal   subject   and
one   from a patient with pulmonary emphysema.   The purpose is to suggest that the observed end point may   not
really   define   the   forced   expiratory vital capacity,
rather the "true" base-line might be lower.   One objective   of our model is to find this true base-line which
would define what we call the "Real FEVC".

attention to the great clinical usefulness of this test.   Of   all
the tests of pulmonary function, the timed vital capacity, now usually referred to generically as the   forced   expiratory   spirogram
(FES)   is   certainly   the most widely used, the most cost/effective
and, when carefully performed, one of the most reliable.   If an individual   has   a perfectly normal forced expiratory spirogram it is
highly improbable that there is any impairment of ventilatory function and unlikely that there is anything wrong with pulmonary function in the overall.   Thus, this simple clinical physiological procedure   deserves   the   close   attention   it has received at so many
hands in the century and more since Hutchinson, and most especially
in   recent years.   There is, indeed, some danger the exquisite simplicity of the forced expiratory spirogram may be buried beneath   a
welter of technically sophisticated analysis.   While it is far from

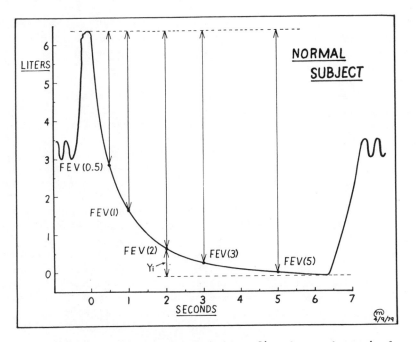

FIGURE 2.   The data needed to fit the mathematical
model   are   the   forced expiratory volumes at 0.5, 1.0,
2.0, 3.0 and 5.0 seconds.   The model considers the   air
yet   to   be   blown out as the prime mover of the forced
expiratory process and these volumes are the Yi   values
obtained   by subtracting the observed forced expiratory
volumes from the current value of FEVC.

a simple matter, it is also not impossibly complicated.   Relatively
elementary   mathematical   models based upon two reasonable premises
can reduce all the information in the   forced   expiratory   spirogram
to just three numbers.   It is the purpose of this report to explore
the potential usefulness of such models.

    The developments here to be reported have, as is most   usually
the   case,   come   to   their   present   state in a number of distinct
steps.   Initially a mathematical model   was   fitted   to   the   first
three seconds of the forced expiratory spirogram using the observed
value of FEVC and the forced expiratory volumes of 0.5,   1.0,   2.0
and   3.0   seconds.   Next,   it was conceived that the observed FEVC
might not be the best or "real" FEVC.   So, as a starting point,   we
took the 5 second FEV as FEVC and incremented it iteratively to see
if the r**2 values as obtained earlier coul be "improved" (FIG. 2).
This   proved   to be the case and very, very close fits to the first
three seconds of the FES could be obtained, with r**2 values

seldom less than 0.998 and often as high as 0.9999.   We conceived
that the FEVC so obtained was the FEVC which "belonged" to the
first three seconds of FES ("the real FEVC") because it emerged
from a model which seemed to describe the early part of the forced
expiratory "process" with remarkable precision (17, 18).   At this
point, we realized we were throwing useful information away by not
fitting to the 5 second FEV as well. This we proceeded to do. The
fits were almost as good as those to the first three seconds and
more of the spirogram "participated" in the modeling.   It is our
conviction that the shape of the spirogram much beyond 5 seconds
involves processes which have nothing to do with normal breathing
or even forced hyperventilation because the subject invokes various
"squeezing" maneuvers to force out the last few milliliters of air
of which he is capable. Therefore, we have continued to address
our attention to the shape of the FES up to five seconds.   Among
normals, 90% of FEVC is usually expelled in 2.0-2.5 seconds. Even
in rather sever grades of pulmonary emphysema and with the value
for "real FEVC" usually larger than "observed FEVC" and thus with
more of the slow part of FES included, 90% of the FEVC is usually
expelled in six seconds and only occasionally does it require over
seven seconds. In our view, the FES beyond five seconds is thus
not likely to contribute much to out understanding of the "process"
of forced expiration.

The two most commonly employed quantitative measures of the
forced expiratory spirogram (FES) are the total magnitude, that is,
the Forced Expiratory Vital Capacity (FEVC), and the Forced Expiratory Volume (FEV) in the first one-half second or the first one
second of the test, FEV(0.5) or FEV (1.0). Both the FEVC and the
FEV may be compared with predicted normal values, referenced most
usually by age, sex and height of the subject and expressed as percent of predicted. The FEV may also be expressed as a percentage
of the observed FEVC. Beyond these there are many other measures
applicable, all of which have certain points to recommend them and
certain defects as well (5). Discussion here is confined to problems experienced with FEVC and with FEV (0.5) or FEV (1.0). There
is little error in measuring the volume attained upon inspiration
before the maximal expiratory effort begins and little error in
measuring the volume at the "end" of the FES, but there is a great
deal of uncertainty about the times where FES begins and ends. The
end really is determined by mutual consent of the technician recording the spirogram and the subject performing the maximal forced
expiratory maneuver. Motivation of both parties has significant
effects.   This is especially obvious in certain patients with pulmonary emphysema who are urged to make a prolonged expiratory effort to try to wring out the last few milliliters of air. For example, our subject .003 was still expiring at a rate of 0.1 liters
per second after 12 seconds of effort at which point he "gave up"
or the technician signaled "that's enough".   More important, the
tracing had become undulatory, indicating that miscellaneous

"squeezing" maneuvers were being employed in an effort to force out still more air. While the air flow was centainly modulated by some intrapulmonary processes, the performance at this point has little to do with normal ventilation or even useful forced hyperventilation.

The uncertainty at the start of forced expiration, where the spirogram "bends over" and becomes relatively steep is commonly dealt with by back extra-polation of a line tangent to the steepest part of the curve to intersect a horizontal line at maximal inspiratory volume thus establishing a point which is taken to be both time zero and volume zero for taking off the various measurements desired. Time uncertainty contributes more error to FEV (0.5) than to FEV (1.0) for the obvious reason that it is a larger fraction of the former than of the latter. These points have been the subject of discussion in recent years (6, 19-22)

These uncertainties, both in time and in volume, suggest that a closely fitted mathematical model based upon five points taken off the spirogram might improve the precision and perhaps the accuracy of measurements derived from the spirogram. For single measurements dependent upon a precise determination of "time zero" there will always be a built-in factor dependent upon the exact technique used to select the take-off point. On the other hand, a model fitted to a curve will, to some extent at least, be self-correcting in this regard if the fit to the curve is truly a close one.

Maximal forced expiration is viewed as a coherent process. Clearly, it occurs in a very complex physical system but it is conceived that it may have regularities such that a relatively simple mathematical model based upon reasonable premises might characterize the process in a useful way. Such has proved to be the case. To see how the equation is derived, it is best to put it into a more tractable form. Consider this model of the FES:

$$\text{Log Log } (FEVC/Y_i) = B + S \text{ Log } T_i$$

Where:

FEVC = A volume systematically related to the process which could (in theory) all be blown out, Real FEVC".

Y_i = Volumes of air still to be expired at times T_i. These volumes are obtained by subtracting the observed volumes from FEVC (Fig. 2).

B = The intercept obtained by linear regression entered with the transformed values of $FEVC/Y_i$ and of T_i.

S = The slope in linear regression.

If equation (1) is a valid model of the forced expiratory process, then so is:

$$\ln \ln (FEVC/Y_i) = B + S \ln Ti \qquad (2)$$

although the values of B and S will be different. Equation (2) can be rewritten:

$$\ln \ln (FEVC/Y_i) = B \ln e + S \ln T_i$$

$$\ln \ln (FEVC/Y_i) = \ln e^{BT_i^S}$$

$$\ln (FEVC/Y_i) = e^{BT_i^S}$$

Let the constant $e^B = b$, then:

$$\ln (FEVC/Y_i) = bT_i^S$$

$$FEVC/Y_i = e^{bT_i^S}$$

$$Y_i = FEVC \cdot e^{-bT_i^S} \qquad (3)$$

Since Y_i is a function of T and FEVC is Y at T = 0, that is, FEVC = Y_o, therefore:

$$Y(T) = Y_o \cdot e^{-bT^S} \qquad \text{The model in exponential form.} \qquad (4)$$

Now the two premises from which it is possible to derive equation (4) are, first, that the rate at which the volume of air in the lungs decreases with time is primarily a function of the volume still to be expelled, and, second, that the rate at which the volume of air in the lungs decreases does itself decrease with time.

These are not exactly new thoughts. The former goes back at least
to Isaac Newton and his first fluxion, while the latter concept
that a rate may be proportional, for example, to the reciprocal of
time, is in the literature of chemistry and physics in places far
too numerous to mention. Take the second premise first and let
$g(t)$ represent some function of time to be defined:

$$\frac{d\ k(t)}{dt} = g(t) \cdot k(t)$$

Let, as proposed, $g(t)$ be inversely proportional to time:

$$g(t) = \frac{a'}{t}$$

Rearrange:

$$\frac{d\ k(t)}{k(t)} = a' \frac{dt}{t}$$

Integrate both sides:

$$\ln k(t) = a' \ln t + \ln a''$$

where $\ln a''$ is the constant of integration. This can be
shown to be equivalent to:

$$k(t) = a'' t^{a'} \tag{5}$$

The first premise was that the rate at which air is exhaled
from the lung is proportional to the amount remaining in
the lung. The proportionality function is $k(t)$, so:

$$\frac{d\ y(t)}{dt} = -k(t) \cdot y(t)$$

where $y(t)$ represents volume.   Rearranging:

$$\frac{d\ y(t)}{y(t)} = -k(t)dt$$

But, from (5) above, $k(t) = a''t^{a'}$, so:

$$\frac{d\ y(t)}{y(t)} = -a''t^{a'}dt$$

Integrate:

$$\ln y(t) = -\ \frac{a''}{a' + 1}\ \cdot\ t^{a'+1}\ + \ln y_0$$

Let $s = a'+1$ and let $b = a''/(a'+1)$, rearrange and take the exponential of both sides:

$$y(t) = y_0 \cdot e^{-bt^s}$$

Equation (6), then, is identical with equation (4).

The only novelty in this matter is the application of this sort of mathematical model to the forced expiratory spirogram. It has obvious similarities to the Gompertz function (23, 24) but, so far, Gompertz functions have not fitted the data nearly as well as the model offered. A model proposed by P. R. J. Burch (25) for the age distribution of radiation induced cancer is quite the same as the one here described:

$$Pr,t = S_0(1 - e^{-k_r t^r})$$

except, of course, that this would correspond to the model herein described subtracted from its equilibrium value. corresponding to the FES recorded as volume exhaled.

To solve the equation:

$$\text{Log Log } (FEVC/Y_i) = B + S \text{ Log } T_i$$

for FEVC, B and S, assume an initial value for FEVC = FEV(5.0) + 0.005, enter the five values of FEVC-FEV and corresponding times T_i, do the transforms, perform linear regression, then find $r**2$, the coefficient of determination. Increment FEVC, repeat the above steps until a new value of r**2 is less than the previous one. Step back in the iteration to the FEVCS which yeilded the best r**2 and calculated B and S.

The original equation:

$$\text{Log Log } (YV/Y_i) = B + S \text{ Log T}$$

which was reported earlier (17, 18) may be derived in essentially the same manner. Generally, it provides a slightly closer fit to FES but a computer program for it runs much more slowly. This is of no concern with a substantial computer such as the DEC PDP 11/40 and, even with a Commodore Pet program in BASIC, run time is only one to three minutes. However, with a programmable calculator such as t the TI-59, the original equation may require an hour or more to search out the "real FEVC". This latter is obviously impractical for routine use.

The equations we have worked with are expressed in Briggsian logarithms for no better reason than that earlier work with this and previous similar mathematical models (26-30) was done with tables of logarithms and be graphic approximations using Log Log paper. Napierian logarithms are obviously more rational and with presently available computing devices there is every reason for using them. We had a large back-log of preliminary work on this model with logarithms to the base 10, so, to the present, we have stayed with these. Anyone who choses to explore the application of this family of models would be well advised to start with natural logarithms and we are in process of such a change-over.

The first one-tenth of a second or so of an actual spirogram is curvilinear and convex upward in contrast to the remainder of the spirogram which is concave upward. Because of this, there has been as earlier noted, much debate as to how to time the forced expiratory volumes (6, 19-22). As a practical matter, in our work we have used the back extrapolation method to set a nominal time zero. The effect of this is to imply a higher flow rate at the beginning of the spirogram than occurred in real life.

There is an important difference between the model proposed in this report and the "true" spirogram, even as modified by the extrapolation method to find time zero, namely, that the very early part of the curve generated by the model is steeper than the actual spirogram even though the latter has been artificially "steepened" by back extrapolation. For example, in our test case .101, at 0.1 seconds the flow rate (from dy/dt) is a reasonable 8.2 liters per second whereas at time 10**-10 seconds, dy/dt gives an absurd 359

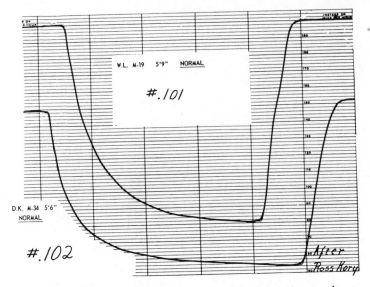

FIGURE 3. These are two normal forced expiratory
spirograms from a file of teaching material put together by Ross Kory some years ago.

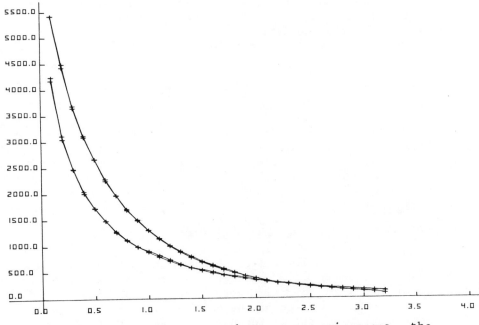

FIGURE 4. Here are those same spirograms, the
plot is the observed data and the spirograms calculated
from the model.

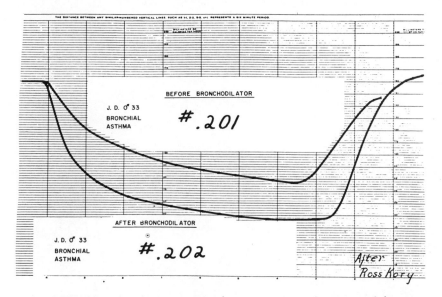

FIGURE 5. These two spirograms are from a subject
with bronchial asthma before and after the administration of bronchodilators.

liters per second. This matter is further discussed where we take
up the relation of the model to the flow volume loop. At this
point, suffice it to say that the model predicts excessively large
forced expiratory volumes in the first few tenths of a second of
the FES. However, since both curves are steep in this region, when
both are plotted together in two dimensions, the curves are seen to
be very close to each other (Figs. 3-6).

After the first few tenths of a second, the values for the
forced expiratory volumes calculated from the model and the observed values are very close. For example, in the case of FEV(0.5)
in our fourteen "test cases", the mean (algebraic) difference
between observed and calculated was -0.008 liters. The average of
the absolute magnitudes of the fourteen "errors" was 0.018 liters
with a standard error (uncorrected for sample size) of +-0.021
liters. In comparison, the original model (YV) yeilded a mean
(algebraic) difference of -0.004, and an average absolute difference of 0.016+-0.019. In the case of the FEV(1.0), the corresponding figures for the new model are 0.022, 0.031 and +-0.039 and for
the original model, 0.014, 0.026 and +-0.032. As earlier stated
and as seen here, the original model gives a slightly better fit
than the new one but in both cases, the differences between the observed and the calculated values are of the same order of magnitude
as the "reading error" in taking the data off a spirographic record. In each of the fourteen "standard test cases" and whether

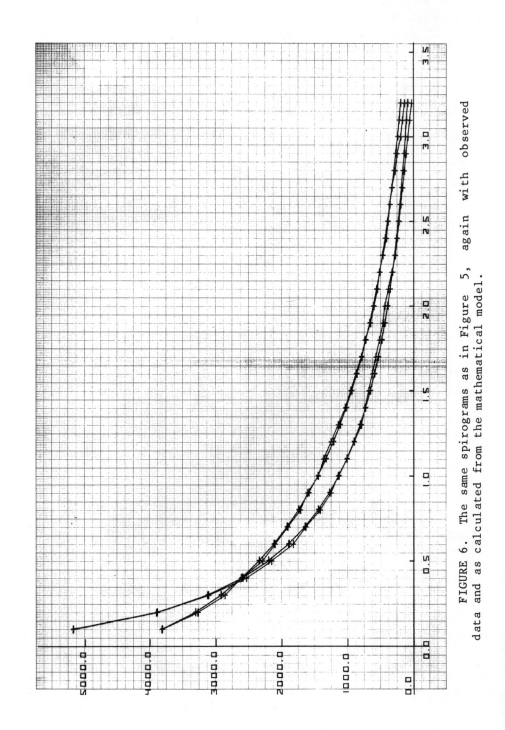

FIGURE 6. The same spirograms as in Figure 5, again with observed data and as calculated from the mathematical model.

for the original or the new model, there are five data points entered for each. These are the FEV values at 0.5, 1.0, 2.0, 3.0 and 5.0 seconds. No significant enchancement of fit is gained by taking more points with either model. In the overall then, there are six differences between calculated and observed values in each case because the five entered values can be compared with the calculated ones and the calculated FEVC compared with the observed FEVC. Note that the latter, in the case of the new model, is FEV(5.0) plus the "FEVC increment" Which produced the best r**2 value. Analysis of these 84 differences between observed and calculated FEV values yields a standard error of the estimate of +-0.018 liters for the original model and +-0.020 liters for the new model. As to differences between calculated and observed values, or "errors" if one will, it is probably true that the calculated value is better than the observed. The graticule of some spirometer recording paper in common use works out to arround 0.040 to 0.050 liters per millimeter and it is difficult to read much more closely than to half a millimeter. Also, a data point calculated from the model is actually based upon a fit to five points.

In pulmonary emphysema, the expiratory process is slowed, so much so, as noted elsewhere, that substantial air flow is usually still in progress when the subject "gives up" the expiratory effort. In contrast, in normal subjects, 90% of the expired air is delivered in less than 2 seconds. Normal standards are, of course, developed from subjects considered to be normal. Therefore, when one compares the magnitude of a forced expiratory vital capacity of an emphysematous subject with a normal standard, one is taking no account of the differences in the time domain. It is possible that the "real FEVC" computed as described may be a more valid volume to compare with a normal standard in that it is a volume inherently coupled to the actual expiratory process of the patient. (FIG. 1)

The entire upslope, the peak and the first portion of the down stroke of the expiratory part of the flow/volume loop is imbedded in the first few tenths of a second of a forced expiratory spirogram. The paper speed as well as the volume scale of most spirometers are such that it is difficult or impossible to take off the early part of the expiratory flow/volume diagram from a volumetric spirographic tracing. The accuracy and precision with which a volumetric spirometer records this part of the spirogram is, of course, a function of the time constant of the device, the accuracy of the clock or timing device, etc., but it is certainly an inappropriate device for accurate and precise recording of the early part of flow/volume diagrams. It is true however, that the effort independent part of the flow/volum diagram can be obtained with precision from the time/volumetric data.

Once the three parameters FEVC, B and S of the model:

$$\text{Log Log } (\text{FEVC}/Y_i) = B + S \text{ Log } T_i$$

have been found, it is a simple matter to derive a "flow/volume" diagram. First, for the abcissa, one generates volumes Vi at 0.1 liter intervals from 0.1 liters out to FEVC. Each of these must be subtracted from FEVC to obtain the corresponding Yi values. For each of these, the corresponding Ti times are calculated from:

$$T_i = 10^{(\text{Log Log }(\text{FEVC}/Y_i) - B)/S}$$

The flow corresponding to each Vi is then calculated from the first differential of:

$$Y_i = \text{FEVC} \cdot 10^{-10^B T_i{}^S}$$

which is:

$$\frac{d(Y_i)}{dT} = -10^B \cdot \ln 10 \cdot \text{FEVC} \cdot S \cdot 10^{-10^B T_i{}^S} \cdot T_i{}^{S-1}$$

actly identical to the effort independent part of the flow sensor derived flow/volume diagram (Figs. 8,9). It is evident that a second differential is equally easy to compute. An analysis of the sort proposed by Mead (32) of SR-VCf curves can be derived.

It would be equally easy to perform transit time analysis (33) upon a spirogram since the exact slope of it is faithfully reproduced by the model, given the three parameters FEVC, B and S. One might speculate upon the value of such an analysis since the early part of the forced expiratory spirogram, even as recorded, but more especially as conventionally measured, has no real relationship to the flow rates moved in the first one or two tenths of a second (FIG. 9). Experience with flow sensitive sensors has revealed that this early part of the flow/volume curve is highly variable as to flow rate although a spirometer reliably captures the gas volumes expired and, later, faithfully reproduces the time course of volume changes once these are within the frequency response characteristic of the instrument.

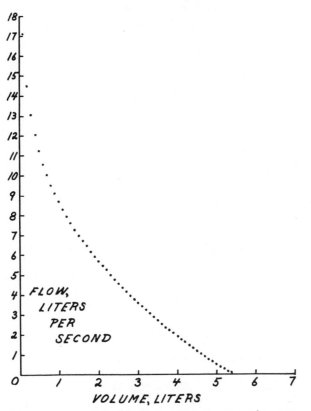

FIGURE 7. A calculation will yield this sort of a
flow/volume diagram. Obviously, the early part of it
has no similarity to the early part of the expiratory
part of the flow/volume loop recorded with flow sensitive sensors.

Then, each value for flow is plotted against the corresponding Vi
value (FIG. 7). The early part of this curve is, of course, quite
dissimilar to the early part of the forced expiratory flow/volume
diagram recorded with flow sensors (31), but the latter part is ex-

Once all the information in the first five seconds of the FES
is condensed into the three parameters FEVC, B and S of equation
(1) it is simple to calculate any of the volumes, flow rates or ratios commonly used to report the FES. These calculated values are
more reliable than those derived from only one or two data points.
The required iterative calculations are possible with programmable
desk calculators but rather slow. A microcomputer as small as the
Commodore PET 8K produces output usually in one to two minutes, the
DEC PDP 11/40 minicomputer in one to three seconds.

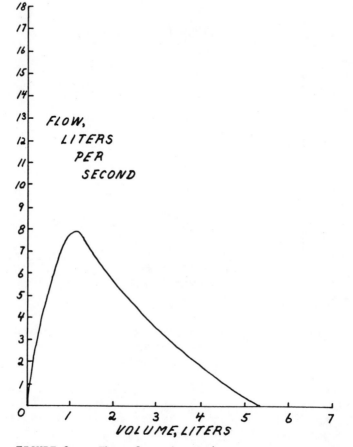

FIGURE 8. The forced expiratory part of a flow/volume loop might look something like this. The relation to the calculated flow/volume diagram may be seen in Figure 9.

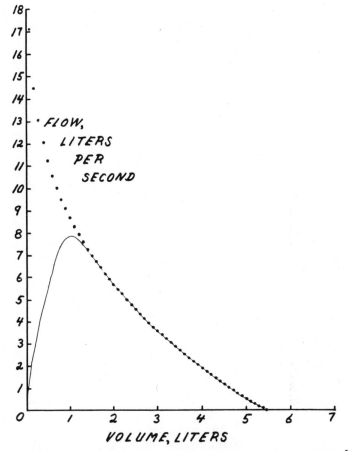

FIGURE 9. Volumetric spirometers are generally
incapable of a faithful rendering of the first few
tenths of a second flow in the forced expiratory spirogram because the response time constants are too slow,
however, after a few tenths of a second, even a lethar-
gic volumetric spirometer is capable of following flow
quite faithfully throughout the effort independent part
of the flow/volume relationship. It is evident the
flow sensor device must give the same answer for flow
versus volume after a second or so as the volumetric
device.

The parameters "Real" FEVC, B and S show promise as useful clinical numbers, however, we lack sufficient data to evaluate them critically at this time. What has been done effectively now is to automate a conventional diagnositc formulation, specifically the quadrant scheme of Miller and Wu (2, 5) as modified by Meneely and Tysinger (3), and, more recently, by Meneely, Light and George (FIG. 10). From this a program has been developed with a simple logic tree to prepare a computer output sheet (FIG. 12) for a patient's chart which reports the usual numerical values derivative

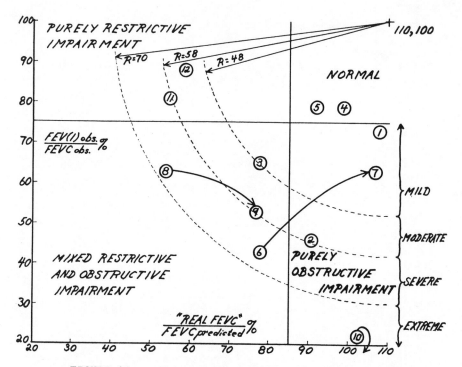

FIGURE 10. The quadrant diagnostic schema of Miller and Wu (2, 3, 5) modified as to the degree of severity of ventilatory impairment and as to the manner in which it is evaluated. Given such a schema, it is easy to see how a computer can be programmed to evaluate a forced expiratory spirogram. Twelve "test spirograms" are plotted. Examples (4) and (5) are illustrated in Figs. 3 and 4; example (6) is bronchial asthma before and (7) after bronchodilators, see Figs. 5 and 6. Figs. 11, 12, 13 and 14 show the computer program applied to example (1).

from FES and which offers an English language expository prose
statement about the kind and the severity of the patient's ventilatory impairment, if any, which can be concluded from the data and
comparison with normal standards. Williams and Brenowitz (34) advised against the use of the term "prohibitive" to characterize the
most extreme degrees of impairment of the forced expiratory spirogram as used originally by Miller and Wu (2) and later by Meneely

```
    RUN SPIRO

DATE:    21-MAY-79 FORCED EXPIRATORY SPIROGRAM

    NAME?            VAH SUBJECT #1

    UNIT.LAB NO.?    79123456.001

    AGE?             20

    SEX? (F=0,M=1) 1

    HEIGHT? (IN)     72


ENTER THE REQUESTED DATA AS MLS OR AS LITERS

    FEV 0.5?    3.54

    FEV 1.0?    4.8

    FEV 2.0?    5.705

    FEV 3.0?    6.08

    FEV 5.0?    6.349

PARAMETER VALUES:

    B (INTERCEPT) = -0.239171
    S (SLOPE) =      0.714874
    R SQUARED =      0.999247
    FEVC INCREMENT     0.095




    DO YOU WANT ANOTHER RUN (Y OR N)?N
    TT3  --  STOP
    >PIP TI:=SPIRO.DAT,LAB.DAT,GRAPH.DAT
```

FIGURE 11. This shows a conversation between an
operator and the program SPIRO in which the necessary
information is put into the computer to fit the model
and run the evaluation. The three output files, SPIRO.
DAT, LAB. DAT and GRAPH. DAT are adapted for output
on a typewriter terminal since these are far more abundant than line printers or graphic printers.

```
DATE:    21-MAY-79  FORCED EXPIRATORY SPIROGRAM

   NAME            VAH SUBJECT #1

   UNIT.LAB NO.    79123456.001

   AGE             20 YEARS

   SEX             MALE

   HEIGHT          72 INCHES

FORCED EXPIRATORY VOLUMES:

   VITAL CAPACITY ("REAL FEVC")        =  6.444    LITERS

   PERCENT OF PREDICTED FEVC           =  109.     PERCENT

   ONE SECOND FEV, FEV(1.0)            =  4.735    LITERS

   PERCENT OF PREDICTED FEV(1.0)       =  100.     PERCENT

   PERCENT OF CALCULATED REAL FEVC     =  73.      PERCENT

FORCED EXPIRATORY FLOW RATES:

   0.1 SECOND FORCED EXPIRATORY FLOW   = 14.552    LITERS PER SECOND

   CANDER AND COMROE, FEF(200-1200)    = 14.600    LITERS PER SECOND

   MID-HALF FLOW RATE, FEF(25%-75%)    =  3.411    LITERS PER SECOND

   PERCENT OF PREDICTED FEF(25%-75%)   =  68.      PERCENT

   75%-85% FLOW RATE, FEF(25%-75%)     =  1.101    LITERS PER SECOND

CLINICAL REPORT OF SPIROGRAM:

   EVALUATION OF THE SPIROGRAM OF VAH SUBJECT #1

   RECORDED ON 21-MAY-79 SHOWS IMPAIRMENT OF VENTILATORY FUNCTION

   WHICH IS MILD IN DEGREE.

   THE CHARACTER OF THE IMPAIRMENT IS PURELY OBSTRUCTIVE.

   THE MID-HALF FORCED EXPIRATORY FLOW RATE IS MILDLY REDUCED WHEN

   COMPARED WITH A PREDICTED NORMAL VALUE OF  4.997  LITERS PER SECOND

   FOR A 72 INCH TALL 20 YEAR OLD MALE.


                  SIGNATURE: ...............................M. D.

                  DATE OF EVALUATION: .....................

CODE:    79123456.001, 20, 1, 72, 6.444, -0.23917, 0.71487, 0.99925, 0.095, 2111
```

FIGURE 12. This is a photograph of SPIRO.DAT, an output sheet on 8 1/2 by 11 inch paper for entry into the patient's record after it has been read, approved and signed by a physician.   Note in addition to the volumes, flows and percentages there is an english language discussion of the spirogram written by the computer from a simple logic tree based upon the Miller and Wu quadrant schema and our own evaluation of severity.

```
SPIROGRAPHIC LABORATORY DATA:

    PROGRAM FOR:  LOG LOG ( FEVC/Y) = B + S LOG ( T)

    DATE:    21-MAY-79    NAME: VAH SUBJECT #1

    UNIT-LABORATORY NO. 79123456.001
    AGE, YEARS            20
    SEX                  MALE
    HEIGHT, INCHES        72


PARAMETER VALUES:

    FEVC =               6.444
    B (INTERCEPT) = -0.239171
    S (SLOPE) =       0.714874
    R SQUARED =       0.999247
    FEVC INCREMENT    0.095


COMPARISON OF OBSERVED WITH CALCULATED VALUES:

    DATUM                  OBSERVED CALCULATED    ERROR

    REAL VITAL CAPACITY      6.444     6.444     0.000    LITERS
    HALF SECOND FEV          3.540     3.574    -0.034    LITERS
    ONE SECOND FEV           4.800     4.735     0.065    LITERS
    TWO SECOND FEV           5.705     5.715    -0.010    LITERS
    THREE SECOND FEV         6.080     6.094    -0.014    LITERS
    FIVE SECOND FEV          6.349     6.347     0.002    LITERS

    STANDARD ERROR OF THE ESTIMATE      +/-     0.033    LITERS


FORCED EXPIRATORY VOLUMES:

    VITAL CAPACITY ("REAL FEVC")     =   6.444   LITERS
    PERCENT OF PREDICTED FEVC        =   109.    PERCENT
    ONE SECOND FEV, FEV( 1)          =   4.735   LITERS
    PERCENT OF PREDICTED FEV( 1)     =   100.    PERCENT
    PERCENT OF CALCULATED REAL FEVC  =   73.     PERCENT


FORCED EXPIRATORY FLOW RATES:

    0.1 SECOND FORCED EXPIRATORY FLOW  = 14.552   LITERS PER SECOND
    CANDER AND COMROE, FEF(200-1200)   = 14.600   LITERS PER SECOND
    MID-HALF FLOW RATE, FEF(25%-75%)   =  3.411   LITERS PER SECOND
    PERCENT OF PREDICTED FEF(25%-75%)  =  68.     PERCENT
    75%-85% FLOW RATE, FEF(75%-85%)    =  1.101   LITERS PER SECOND


ENGLISH LANGUAGE DIAGNOSIS IN CODE               2111.
```

FIGURE 13. The second file prepared by the computer,
LAB.DAT, of which this is a photograph, is for laboratory
reference purposes. Note the comparison of calculated with
observed values in the middle of the page.

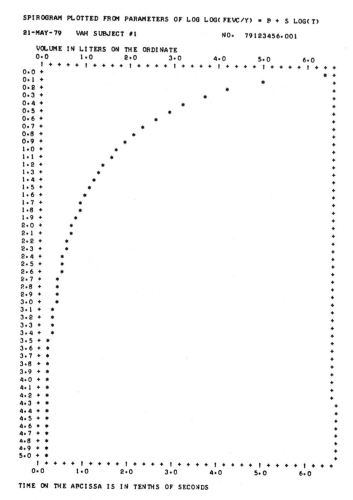

SPIROGRAM PLOTTED FROM PARAMETERS OF LOG LOG(FEVC/Y) = B + S LOG(T)

21-MAY-79   VAH SUBJECT #1   NO. 79123456.001

FIGURE 14. This is a photograph of the third file,
GRAPH.DAT, a graphic prepared for typewriter terminals which
more or less reproduces the spirogram. The original spirogram
must, for legal reasons, remain in the laboratory where the
test was performed but there are times when a physician seeing
a consultation case might wish to send a copy of the spirogram
along with his report to the referring physician.

and Ferguson (3) because, as they showed in 16 patients, such impairment was indeed not prohibitive of major surgery. Miller, Paez and Johnson no longer use the term (5) and we ahve picked up their "extreme" as well. Figure 13 presents a data sheet for laboratory research purposes. If one wishes, the spirogram can be reproduced using computer graphics at any level of sophistication available to the operator. Figure 14 shows such a presentation, designed for typewriter terminals since these are much more abundant than printers with real graphics capability. As small a resource as the TI-59 can prepare useful output, but as noted above, it is slow. Five to eight minutes are required to obtain the parameters FEVC, B and S. Then, if the subject is normal, another two minutes or, if impaired, three minutes for a total time of seven to thirteen minutes. On the other hand, it is uncertain one could calculate and record a like amount of information manually in any less time. Figure 15 presents four examples of TI-59 output.

Finally, to summarize, we have the original model presented a year ago (17, 18). Since then, we have developed the new model here reported. We lack the data to declare the new model is just as good as the original but it seems so and runs much much faster in the computer. It also makes more sense mathematically. Both can be derived rigorously from just two premises, one, that the rate at which the volume of air in the lung decreases with time in the forced expiratory spirogram is proportional to the volume of air not yet exhaled, and second, that the rate itself diminishes as a function of the inverse of time.

It will take an appreciable length of time to collect a sufficient volume of data to examine the real clinical usefulness of these models. For this reason, it appreared worthwhile to report our progress to date. Whatever else may be said of this model, it rather precisely fits all the sorts of forced expiratory spirograms encountered in clinical practice, it embodies in three numbers all the information in a give spirogram, it appears probable these numbers will have useful clinical correlates, it provides an objective method for obtaining an FEVC which has a formal and specific relationship to all the processes in progress in the first five seconds of the forced expiratory spirogram, and, without any respect to its possible other merits, it provides a means for rapid efficient computer processing of forced expiratory spirographic data into an output format with which the clinician is familiar.

```
UNIT.LAB? 0.402              UNIT.LAB? 0.002              UNIT.LAB? 0.003              UNIT.LAB? 0.303
   FEVC?  2.944                 FEVC?  3.861                 FEVC?  3.5                   FEVC?  1.355
      B?                           B?                           B?                           B?
      -0.035252                    -0.571114                    -0.346004                    -0.718714
      S?                           S?                           S?                           S?
      0.574722                     0.666195                     0.641929                     0.923471
   HEIGHT?  69.                 HEIGHT?  69.                 HEIGHT?  69.                 HEIGHT?  69.
   AGE?  39.                    AGE?  69.                    AGE?  59.                    AGE?  43.
   SEX?  M=1, F=0               SEX?  M=1, F=0               SEX?  M=1, F=0               SEX?  M=1, F=0
         1.                           1.                           1.                           1.

FEVC‡      2.944  LITS        FEVC‡      3.861  LITS        FEVC‡      3.5    LITS        FEVC‡      1.355  LITS
% OF PRED.   58.  %           % OF PRED.   90.  %           % OF PRED.   77.  %           % OF PRED.   27.  %
FEV(1)‡    2.592  LITS        FEV(1)‡    1.780  LITS        FEV(1)‡    2.260  LITS        FEV(1)‡    0.482  LITS
% OF PRED.   67.  %           % OF PRED.   61.  %           % OF PRED.   70.  %           % OF PRED.   12.  %
% OF OBS.    88.  %           % OF OBS.    46.  %           % OF OBS.    64.  %           % OF OBS.    35.  %
FEF 25%-75%‡ 3.304 L/S        FEF 25%-75%‡ 0.634 L/S        FEF 25%-75%‡ 1.220 L/S        FEF 25%-75%‡ 0.239 L/S
% OF PRED.   82.  %           % OF PRED.   23.  %           % OF PRED.   39.  %           % OF PRED.    6.  %

SPIROGRAM SHOWS              SPIROGRAM SHOWS              SPIROGRAM SHOWS              SPIROGRAM SHOWS
IMPAIRMENT                  IMPAIRMENT                  IMPAIRMENT                  IMPAIRMENT
WHICH IS                   WHICH IS                   WHICH IS                   WHICH IS
MODERATE                   MODERATE                   MODERATE                   EXTREME
AND WHICH IS               AND WHICH IS               AND WHICH IS               AND WHICH IS
RESTRICTIVE                OBSTRUCTIVE                MIXED                      MIXED
IN CHARACTER.              IN CHARACTER.              IN CHARACTER.              IN CHARACTER.
25%-75% FLOW IS            25%-75% FLOW IS            25%-75% FLOW IS            25%-75% FLOW IS
MILDLY                     SEVERELY                   SEVERELY                   EXTREMELY
REDUCED.                   REDUCED.                   REDUCED.                   REDUCED.
```

FIGURE 15. Never underestimate the power of a Texas Instruments TI-59! These four strips show above the horizontal line the input of the three parameters and the patient ID. Below the line is the output, worked up by a program in the TI-59, which includes and english language evaluation. The three parameters, FEVC, B and S are obtained from another program entered with five values of FEV taken at 0.5, 1.0, 2.0, 3.0 and 5.0 seconds from the recorded forced expiratory spirogram.

BIBLIOGRAPHY

1. MILLS, JN: Variability of the vital capacity of the normal human subject. J. Physiol. 110, 76-82, 1949.

2. MILLER, WF, WU, N AND JOHNSON, RL, JR.: Convenient method of evaluating pulmonary ventilatory function with a single breath test. Anesthesiology 17, 480-493, 1956.

3. MENEELY, GR AND FERGUSON, JL: Pulmonary evaluation and risk in patient preparation for anesthesia and surgery. J. Am. Med. Assoc. 175, 1074-1080. 1961

4. COMROE, JH JR, FORSTER, RE, DU BOIS, AB, BRISCOE, WA AND CARL-SON, E: The Lung, Clinical Physiology and Pulmonary Function Tests. 2nd Ed., The Year Book Publishers, Chicago, 1962.

5. MILLER, WF, PAEZ, PN AND JOHNSON, C: Laboratory evaluation of pulmonary function. Chapter 3 in Laboratory Medicine, Vol. III, G. J. Race, Ed., 3rd revision, Harper and Row, New York, 1977.

6. Report of the Snowbird Workshop on Standardization of Spirometry, ATS News, 3, 20-25, 1977 and Am. Rev. Resp. Dis., 119, 831-838, 1979, and Standardization of Spirometry (Editorial, Renzetti, A. D.), Ibid, 693-694.

7. DAVY, H.: Researches Chiefly Concerning Nitrous Oxide. J. Johnson, London, 1800.

8. DAVY, H.: Researches, chemical and philosophical, chiefly concerning nitrous oxide, or dephlogisticated nitrous air, and its respiration. London, Smith, Elder and Co., 1839. (in collected works)

9. HUTCHINSON, JOHN: On the capacity of the lungs, and on the respiratory functions, with a view of extablishing a precise and easy method of detecting disease by the spirometer. Med. Chir. Soc. Trans. XXIX, 1846 pp 137-252

10. BARTELS, H, BUCHERL, E, HERTZ, CW, RODEWOLD, G AND SCHWAB, M.: Methods in Pulmonary Physiology. Trans. by John M. Workman, Hafner Publishing Co., Inc., New York-London, 1963.

11. VOLHARD, F: in discussion of: Bonniger, M. Zur Physiologie und Pathologie der Atmung, Verh. Dtsch. Ges. Inn. Med. 25, 514-529, 1908. Volhard's discussion is on page 530.

12. RAITHER, E: Studien uber Emphysem. Beitr. A. Klin. d. Tuberk. Wurzb. 1912, XXII, 137-164.

13. TIFFENEAU, R. AND PINELLI, A: Air circulant et air captif dans l'exploration de la fonction ventilatrice pulmonaire. Paris Med. 37, 624-628, 1947.

14. TIFFENEAU, R. AND PINELLI, A: Reguation bronchique de la ventilation pulmonaire. J. F. Med. Chir. Thorac., 2, 221- , 1948.

15. GAENSLER, E A: Analysis of the ventilation defect by timed capacity measurements. Am. Rev. Tuberc. 64, 256, 1951.

16. GAENSLER, EA: An instrument for dynamic vital capacity measurements. Science 114, 444, 1951.

17. MENEELY, G.R., KNOX, F.S. III, WALKER, R.F., JR., GEORGE R.B., GIRARD, W.M. and MATTSON, D.A. Will the real FEVC please stand up? A computer analysis. Proceedings of the Assoc. for the Adv. of Med. Instrum., 13th Annual Meeting, March 29-April 1, 1978, p. 193.

18. KNOX, F.S. III, WALKER, R.F., JR., MENEELY, G.R., CONRAD, B.S., GIRARD, W.M. and GEORGE, R.B. Automated clinical diagnosis of the forced expiratory spirogram, Ibid., p. 155.

19. KORY, R.C., CALLAHAN, R., BOREN, H.D. and SYNER, J.C. The Veterans Administration cooperative study of pulmonary function. I. Clinical spirometry in normal men. Am. J. Med. 30, 243, 1961.

20. KAUMER, R.E. and MORRIS, A.H. Clinical Pulmonary Function Testing, Instrumentation Thoracic Society, Salt Lake City, Utah, 1975, p. 1.

21. SMITH, A.A. and GAENSLER, E.A. Timing of forced expiratory volume in one second. Am. Rev. Respir. Dis. 112, 882-885, 1975.

22. KNUDSON, R.J., LEBOWITZ, M.D. and SLATIN, R.C. The timing of the forced vital capacity. Am. Rev. Respir. Dis. 119, 315-318, 1979.

23. GOMPERTZ, BENJAMIN (1779-1865), English mathematician and actuary, Phil. Trans. R. Soc. 115, 513-585, 1825.

24. NORTON, L., SIMON, R., BRERETON, H.D. and BOGDEN, A.E. Predicting the course of Gompertzian growth. Nature 264, 542-545, 1976.

25. BURCH, P.R.J. Natural and Radiation Carcinogenesis in Man. I. Theory of initiation phase. Proc. R. Soc. (London) - B. Biological Sciences 162, 223-287, 1965.

26. MENEELY, G.R. Mixing in biological systems. Proc. So. Soc. Clin. Res., Am. J. Med. 19, 148, 1955.

27. DYBICKI, J., MERRILL, J.M., BALCHUM, O.J. and MENEELY, G.R. Intrapulmonary helium mixing. Fed. Proc. 18, 39, 1959.

28. DOUGLAS, B. and MENEELY, G.R. The rate of absorption of radioactive isotopes as a test of the efficiency of the circulation in pedicle flaps. Plastic and Reconstructive Surgery 20, 350, 1957.

29. MENEELY, G.R., BALCHUM, O.J., MERRILL, J.M., DYBICKI, J., MATTSON, D.H., CARP, D.A. and GEORGE, R.B. A simple equation describing intrapulmonary gas mixing. In Volume in Honor of Thomas Doxiades, Hospital Evangelismos, Athens, 1976, pp. 356-374.

30. LIGHT, R.W., GEORGE R.B., MENEELY, G.R., CHOUSET, L.J., ROSENKRANS, R.H. A new method for analyzing multiple breath nitrogen washout curves. Submitted for publication.

31. HYATT, R.E., SCHILDER, D.P. and FRY, D.L. Relation between maximum expiratory flow and degree of lung inflation. J. Appl. Physiol. 13, 331-336, 1958.

32. MEAD, J. Analysis of the configuration of maximum expiratory flow-volume curves. J. App. Physiol. 44(2), 156-165, 1978.

33. FISH, J., MEUKES, H., ROSENTHAL, R., SUMMER, W., NORMAN, P. and PERMUTT, S. The effect of acute bronchospasm on distrubution of transit times during forced expiration. Am. Rev. Resp. Dis. 109, 700, 1974.

34. WILLIAMS, C.D. and BRENOWITZ, J.B. "Prohibitive" lung function and major surgery. Am. J. Surg. 132, 763-766, 1976.

AUTOMATED INTERPRETATION OF PULMONARY FUNCTION TEST RESULTS

J.C. Kunz, R.J. Fallat,
D.H. McClung, J.J. Osborn

The Institutes of Medical Sciences
Pacific Medical Center
2200 Webster Street
San Francisco, California 94115

B.A. Votteri

Sequoia District Hospital
Alameda & Whipple Streets
Redwood City, California 94062

H.P. Nii, J.S. Aikins,
L.M. Fagan, E.A. Feigenbaum

Department of Computer Science
Stanford University
Stanford, California 94305

We present results of the use of a computerized system for the automated interpretation of standard laboratory measures of pulmonary function and arterial blood gases. This system is now in routine use in Presbyterian Hospital, Pacific Medical Center. The program produces a report, intended for patient records, that explains the clinical significance of measured quantitative test results and gives a diagnosis of the presence and severity of pulmonary disease in terms of the measured data, referral diagnosis, and patient history. "Rules" are used by both the physiologist and the computer system to specify the system operation. "Rules" are statements of the form " IF [condition] THEN [conclusion]", ie "if a set of facts are true" then "make a conclusion based on these facts". The sequence of rules used to interpret the case also specifies a line of reasoning about the case. The rules also contain information usable for making a detailed explanation of the interpretation of the case. The use of rules for this type of knowledge based system is taken from the results of applied Artificial Intelligence research.   In a 144 case prospective evaluation, there was a 91% overall rate of agreement between the rule based system diagnoses and the diagnoses of the designing physiologist;   there was a 89% rate of agreement between the system diagnoses and diagnoses of a second independent physiologist.

PUFF interprets standard laboratory measures of pulmonary function.   Interpretation of these tests involves identifying the presence of disease and determining its severity.   The system recognizes obstructive Airways Disease (OAD: indicated by reduced flow rates during forced exhalation), Restrictive Lung Disease (RLD: indicated by reduced lung volumes), and alveolar-capillary Diffusion Defect (DD: indicated by reduced diffusivity of inhaled CO into the blood), and Neuromuscular disease (NMD: indicated by reduced flow rates during inhalation and reduced lung volumes).   We define four levels of severity for each disease type.   Obstruction and restriction may exist concurrently, and the presence of one mediates some indications of the other.   The system recognizes three subtypes of airway obstruction: asthma, bronchitis, and emphysema.   It also interprets results of tests performed before and after bronchodilation as well as arterial blood gas measurements taken at rest and following exercise.

PUFF is now in routine use in Presbyterian Hospital, Pacific Medical Center (PMC), in San Francisco.   The program produces a report, intended for the patient record, that explains the clinical significance of measured quantitative test results and gives a diagnosis of the presence and severity of pulmonary disease in terms of the measured data, referral diagnosis, and patient history.   Each report is read by an MD.   The staff physicians are free to accept, modify or reject the PUFF interpretations.   Most reports are now accepted without change;   they are signed and entered into the patient record.   The remaining reports usually require simple additional comments about the relation of current to previous test results.   The consistency and detail of the explanation of results has proved to be an attractive service to the staff.

"Rules", or statements of the form "IF [condition] THEN [conclusion]", are used by the physiologist and the computer system to specify the system operation.   The sequence of rules used to interpret the case also specifies a line of reasoning about the case, or the detailed explanation of the interpretation of the case.

In a 144 case prospective evaluation, there was a 91 overall rate of agreement between the rule based system diagnoses and the diagnoses of the designing physiologist;   there was an 89 rate of agreement between the system diagnoses and diagnoses of a second independent physiologist.

The use and implementation of production rules for pulmonary function test interpretation were achieved using techniques of applied Artificial Intelligence research. These techniques provided direct computer manipulation of the clinical knowledge represented symbolically by the physiologist. In addition, the symbolic manipulation allows elucidation of a line of reasoning, or the sequence of judgments used to interpret individual measurements and to form diagnoses based on these judgments. This explainable reasoning process is appropriate for interpreting individual patient cases, whereas an analytical approach, such as statistical discriminant analysis, is more useful for characterizing groups of cases. The explanation of the test results contributes to the physician's understanding of the patient status and this explanation gives credibility to the diagnosis.

DISCUSSION

The PUFF system was developed in the tradition of applied Artificial Intelligence research. The general problem of interpreting the clinical significance of clinical information is difficult for a number of reasons including the essential complexity of the problem and the fact that data has different meanings in different biomedical situations. The biomedical members of our collaborative team had experienced frustration over a long term with the difficulty of dealing with the complexity of clinical data interpretation problems in an effective manner. The object of the interpretation was clear to us: clinical decision making. Clinical decision making was viewed in a traditional way as including diagnosis and recommendation of therapy. We admit to interest in, but we have attempted to avoid, interpretation of physiology for its own sake. Techniques of statistical decision making and Bayesian Analysis we felt inappropriate for pulmonary function test interpretation because they do not suggest direct and effective use of the available physiological understanding of the mechanism of pulmonary disease and the well understood relation between measurement and underlying function. Our attempts at programming "solutions" to the pulmonary function test interpretation problem led us to focus on problems of computer programming, and not on problems of pulmonary physiology and test interpretation.

As the results which is possible with symbolic processing of this paper suggest, we have had success with the rule based approach to the test interpretation process. The process we followed was to identify a set of interpretation rules which would be sufficient for interpreting a large and useful set of test cases. Then we viewed these rules as the data to a computer program, or more specifically as a language which was to be interpreted and manipulated by computer. The vehicle which allowed us to interpret our clinical language was a generalized production rule interpretation program, EMYCIN [vanMelle], which provided a reasonably general mechanism for interpreting production rules and for "chaining" them together in a manner conforming to simple rules of inference (ie modus ponens). In a matter of several weeks, we had made an initial definition of a set of interpretation rules, incorporated them into the EMYCIN system, modified the EMYCIN system to process some of the specialized syntactic and semantic constructs which we had included in the rule set, and ran the system with a small number of test cases.

The effect of this very rapid realization of a test mechanism was very encouraging: we confirmed our working hypothesis that pulmonary function test interpretation was tractable, in principle. In addition, we found the rule representation for the problem to be appropriate for the physician, appropriate for the computer programmers, and acceptable to the computer. In retrospect, the key capability to developing the system laid in the choice of the representation and the accessibility of a set of tools to manipulate the problem with only minimal regard for the computer programming. Having convinced ourselves of the appropriateness of the problem representation, we developed a parallel procedural system on a minicomputer, in BASIC, which was a program to "execute" the production rules. The conversion to the minicomputer and BASIC was made for simple expediencey -- the minicomputer program was much smaller and faster than the EMYCIN-based interpreter. We have maintained the production rule problem specification, however -- the rule set listing defines the interpretation process for both the physician and the computer programmer. The programmer must translate the rules into BASIC, while this function is performed automatically by the automated rule interpreter.

While procedural implementations may be relatively efficient computationally, a limitation of the procedural system implementation is the fact that rule invocation is controlled by the explicit control structure of the computer program (ie with loops, GOTO's and sequential program statement execution). Thus, the designer and implementer must identify and provide this control, and control issues usually are a diversion from the consideration of the physiological and clinical problems of test interpretation. Our experience is that the flexibility of easy rule changes, a flexibility associated with use of the rule interpreter, was critical during the early system work associated with problem definition. Once the problem was well defined, then the relative inflexibility of procedural problem implementation was an acceptable price to pay in return for the computational efficiencey we wanted for the operational system.

REFERENCE

1. VAN MELLE, W., "A Domain Independent Production-Rule System for Consultation Program", Proceedings the International Joint Conference on Artificial Intelligence, 1979.

EXTRACTING INFORMATION FROM BLOOD GAS ANALYSES

John E. Brimm, M.D., Clifford M. Janson,
Maureen A. Knight, and Richard M. Peters, M.D.

Department of Surgery/Cardiothoracic
University of California Medical Center
San Diego, California

ABSTRACT

Incorrect decisions regarding patient management can result
when physicians misinterpret blood gas analyses because they do not
calculate, or do not know how to calculate, important derived values. To assure complete analysis of available data from the patient, we have developed a computer-based system for extracting maximum information from the tests performed. Whenever a patient's arterial and/or venous blood gas data are entered, the system cor-
rects them to body temperature, calculates the O2 saturation, base
excess, O2 content and predicted PaO2 for various FIO2's, and provides an interpretation of acid-base status. It then integrates
the blood gas data with other data collected by our computerized
cardiopulmonary testing system. The system automatically searches
for time-windowed results of various respiratory and hemodynamic
tests--cardiac output determinations, end-tidal gas concentrations
and gas exchange. Then it computes the available oxygen, shunt
fraction, arterial-alveolar oxygen difference, dead space-to-tidal
volume ratio, O2 consumption, and other values.

GLOSSARY

A-a DO2	Alveolar-arterial oxygen gradient
A-V DO2	Arterio-venous oxygen difference
CaO2	Arterial oxygen content
CAO2	Oxygen content of blood draining well ventilated alveoli
CvO2	Oxygen content of mixed venous blood
CPAP	Continuous positive airway pressure
FIO2	Fraction of inspired oxygen
PaCO2	Arterial carbon dioxide partial pressure
PaO2	Arterial oxygen partial pressure
PaH	Arterial pH
Qs/Qt	Intrapulmonary shunt fraction
SaO2	Arterial oxygen saturation
VA/Q	Ventilation-perfusion ratio

Arterial blood gas analyses are used for major control of cardiorespiratory therapy in the ICU. Unfortunately, clinical users often misinterpret blood gas analyses because they do not understand the interactions affecting partial pressures and pH. The important basic elements for interpreting arterial pH, PCO2, and PO2 are the shapes of the oxyhemoglobin and CO2 dissociation curves of blood and the Henderson-Hasselbalch and shunt equations. Knowledgeable clinicians recognize that simplistic analysis, neglecting the full implications of these factors, can lead to erroneous conclusions and harmful treatments.

Because the shapes of the CO2 and O2 dissociation curves differ and because ventilation-perfusion incoordination is the principal dysfunction in pulmonary disease, PaCO2 should be used to adjust the level of alveolar ventilation. The Henderson-Hasselbalch equation and the role of the lungs in controlling blood acidity dictate that the PaCO2 should be assessed on the basis of the hydrogen ion level, not a fixed "normal level". In metabolic acidosis, a "normal" PaCO2 of 40 torr is indicative of severe ventilatory dysfunction.

If oxygen saturations are not available, physicians err in raising the FIO2 in an attempt to improve oxygenation when blood is already fully saturated. Analysis of altered PaO2 is often incorrect because clinicians persist in using A-a DO2 on 100% oxygen for estimating the degree of pulmonary dysfunction. Continued use of this maneuver ignores the fact that an FIO2 of 1.0 leads to the collapse of low VA/Q alveoli (1) and disregards the effects of CvO2 and hemoglobin concentration on CaO2 at the same shunt level. To avoid using 100% inspired oxygen, some clinicians suggest using the A-a DO2: FIO2 ratio. This partial correction neglects the shape of the O2 dissociation curve and the effect of the CvO2.

Such sloppy reasoning has led not only to publications that totally misinterpret A-aDO2 (2), but also to grievous errors in therapy. A patient with a decreased cardiac output, resulting in a low CvO2 and PaO2, might be put on CPAP or PEEP, further lowering the cardiac output. The proper therapy should be to correct the low cardiac output. To assure a complete assessment of the available data from patients, we have developed a computer-based system for extracting the maximum information from the tests performed. This automated system can remove dependence on simplistic calculations and help to avert mistakes by providing interpretations of blood gas data.

Description of the System

Computers have been used for partial interpretation of blood gas data for nearly a decade. Bleich (3) and Goldberg (4) have reported systems in which physicians enter data interactively for consultation in acid-base disorders. Gardner (5) has developed a system bypassing the requirement for physician data entry for acid-base interpretation, classification of oxygen status, and calculation of alveolar-arterial oxygen gradients. Data are captured directly from blood gas instruments, and results are reported by both computer terminals and a telephone voice response system.

Our system also provides an acid-base interpretation but goes considerably further in calculating derived variables based on other physiologic data. If both arterial and venous blood gas values have been measured, for example, then the shunt fraction will be automatically calculated. The system uses a hierarchy of clinical assumptions in order to calculate as many derived variables as possible. The overall system logic is shown in FIG 1.

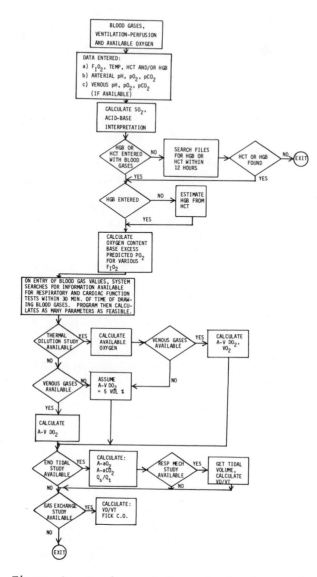

Figure 1.   Logic used for calculating derived variables.

The system is based in the Hewlett-Packard 5600A Patient Data Management System.   Physicians and nurses are not required to enter the needed information;   instead blood gas analyses are entered by technicians using computer terminals in the two blood gas laboratories in our hospial - one run by the Department of Anesthesiology and the other by the Division of Pulmonary Medicine. The computer system performs all possible calculations and interpretations automatically, and then communicates this information to the ICU clinical staff.

Temperature Correction

The system permits entry of both arterial and venous blood gas values (pH, PO2, PCO2) either uncorrected (i.e., at 37°C) or corrected for patients' body temperatures. Newer blood gas machines adjust values to body temperature automatically, but not all machines used in our hospital do. Since the system uses temperature-corrected and uncorrected values at different stages of its calculations, temperature-corrected values are uncorrected and vice-versa. The temperature correction equations of Thomas (6) were inverted for "uncorrection". All reports and interpretations use values corrected to patients' temperatures.

Acid-Base Interpretation

Acid-base interpretation of PaCO2 and PaH uses the method of Burrows (7). The PaH/PaCO2 diagram is divided into 22 regions, most of which are convex. Non-convex regions are divided into convex subregions. Each region is described by an interior point and a set of vertices at the intersection of linear boundaries. The regions are tested in succession to determine whether the patient's pH and PaCO2 fall into the interior. When the proper region is detected, the acid-base interpretation message for that region is given. Base excess and bicarbonate are calculated by Thomas' method (6).

Oxygen Saturation and Content

If the venous or arterial oxygen saturation or content is measured in the laboratory, it may be entered directly. Otherwise the system will try to calculate these values. Oxygen saturation is calculated using Thomas' revision (6) of Kelman's algorithm (8). Oxygen content of blood is the sum of the oxygen bound to the hemoglobin and that dissolved in the blood. Thus, the first term can be calculated as a constant times the saturation times the hemoglobin concentration; the second is the product of the oxygen solubility coefficient and the PO2. If both the arterial and venous gases are entered, then the arterial-venous oxygen difference is calculated as follows: A-V DO2 = CaO2 - CvO2.

HEMOGLOBIN/HEMATOCRIT

The handling of missing hemoglobin and hematocrit values illustrates our attempt to use a hierarchy of clinical assumptions in order to provide information for care. The hemoglobin concentration is needed for calculating the oxygen content, which in turn is used for calculating shunt fraction, one of the primary criteria for adjusting ventilatory therapy. Hematocrit is used in Thomas' equations (6) for calculating the base excess, oxygen saturation, and temperature correction for PO2.

In the clinical situation, however, hemoglobin concentration and hematocrit are typically not both measured every time a blood gas is determined. In the absence of blood loss or transfusion, they simply do not change as rapidly as blood gas values, and frequent measurement is not warranted. A computer algorithm for calculating derived variables must cope with the "time constants" of the primary physiological variables.

If hemoglobin concentration and hematocrit have both been measured, we use both. If only one has been measured, then we estimate the other. (9) This estimation could lead to slight errors in the event of chronic anemia. If neither value has been entered, then the system searches for the most recently measured value within the prior 12 hours. The validity of using this prior value depends on the absence of blood loss or transfusion; however, the presence of either condition will usually result in more frequent determinations. If no value for hemoglobin or hematocrit can be found, then some calculations, such as those for oxygen content and shunt fraction, are omitted. Base excess and the temperature correction for PO2 are still calculated, however, with less precise formulae.

SHUNT FRACTION

The shunt fraction (Qs/Qt) is a measure of the fraction of blood going to unperfused alveoli. It can be used to define anatomically unperfused alveoli or a physiologic equivalent to unperfused alveoli. It is defined by the equation:

$$Q_s/Q_t = \frac{C_AO_2 - CaO_2}{C_AO_2 - CvO_2}$$

where CaO2 and CvO2 are the oxygen contents of arterial and mixed venous blood, and CAO2 is the oxygen content of blood draining well-ventilated alveoli.

Because CAO2 cannot be directly measured, it must be approximated using the alveolar oxygen concentration, which also cannot be

directly measured. If end-tidal gas concentrations have been measured using our on-line respiratory testing system (11), then the
end-tidal concentration is used for calculating CAO2.   Otherwise,
alveolar oxygen is approximated from the FIO2 and PaCO2 (10).
These same assumptions are used for calculating the
alveolar-arterial oxygen and CO2 gradients.

   If venous gases have not been measured, the system still attempts to calculate Qs/Qt using the assumption that the A-V DO2 is
five volumes percent. This assumption holds when the cardiac output is not depressed (10). When cardiac output is low, a Swan-Ganz
catheter should be inserted for measurement of left heart filling
pressure and thermodilution cardiac output, and blood can be sampled to measure venous gas partial pressures. In our experience,
using FIO2 instead of end-tidal O2 and using an assumed A-V DO2
makes little difference in the calculated shunt fraction when the
PaO2 on room air is above 60 torr (10). When the PaO2 is less than
50 torr on room air, or less than 230 torr on an FIO2 of 1, A-V DO2
should be measured.

Predicted SaO2 and PaO2 for Various FIO2's

   This part of the program informs the user of the increase in
PaO2 that can be expected from a given increase in FIO2. If one
assumes that the shunt fraction and A-V DO2 do not change, the
shunt equation can be used to predict the effect of altering FIO2
on PaO2 and oxygen saturation (SaO2).

The shunt equation can be rewritten as:

$$Q_s/Q_t = \frac{C_AO_2 - CaO_2}{C_AO_2 - CaO_2 + \text{A-V } DO_2}$$

This equation can be rearranged to:

$$CaO_2 = C_AO_2 - \frac{\text{A-V } DO_2 \times Q_s/Q_t}{(1 - Q_s/Q_t)}$$

Under our assumptions the second term on the right is a constant.
CAO2 is approximated from the FIO2. Consequently, since with Thomas' equations (6) the SaO2 can be derived for a given PaO2, pH,
and base excess, then the CaO2 defines a unique PaO2, SaO2 combination for a given hemoglobin concentration. The interpretation program reports these PaO2, SaO2 combinations for FIO2's of 0.21 (room
air), 0.3, 0.4, 0.5, 0.6, 0.8, and 1.00 (100% oxygen), as shown in
FIG 2.

```
    121111     TEST ME                    ANAES

ARTERIAL BLOOD GASES

PAO2 & %SAT PREDICTED AT VARIOUS FIO2'S:
FIO2   .21   .3    .4    .5    .6    .8    1.0
       ----  ----  ----  ****  ----  ----  ----
PAO2    68    78    88   100   119   188   302
SAT.    92    95    96    97    98    99    99
        A PAO2 OF AT LEAST 65 MM HG CAN BE
ACHIEVED ON ROOM AIR.
        IF PAO2 DEPRESSION IS ALL DUE TO
SHUNT WITH FIO2 OF 1, A-AO2 WOULD BE 358
(FOR FIO2 > .5, THIS IS A GOOD APPROX.)


PRESS GO TO CONTINUE OR CHOOSE "9" TO
REVIEW THIS STUDY, OR CHOOSE "10" TO
REVIEW VENOUS BLOOD GASES :
```

Figure 2. Computer display used for reporting
PaO2 and SaO2 calculated for various FIO2's assuming a
constant shunt fraction and A-V DO2. The patient's
current FIO2 is denoted with asterisks.

Integration With Other Physiologic Data

Each of the calculations mentioned thus far is made using arterial blood gases, sometimes in combination with venous blood
gases, FIO2, hemoglobin and hematocrit, with the exception that the
shunt fraction is better estimated if the end-tidal oxygen concentration is measured. The system then attempts to integrate blood
gas data with other results obtained from our automated cardiopulmonary testing capability (11). The system searches for results of
thermodilution cardiac output determinations, end-tidal gas concentrations, and tidal volume and gas exchange measurements. Since
these measurements are almost never done simultaneously in the ICU
setting, the system uses the criterion that they be done at approximately the same time. Approximately is defined to be within 45
minutes prior to or 15 minutes following the blood gas determination. The size of the time window has been established to permit
as much data as possible to be integrated without disregarding the
physiologic variability of the measurements. The clinical rationale for the offset time window is that measurements are made after
enough time has elapsed to permit equilibration to a prior therapeutic maneuver and just before a subsequent maneuver. Because
blood gases may be entered before some of the results from the
on-line testing system, but within the time window, entry of any of
the other primary physiologic variables also triggers the full calculation system.

If cardiac output and A-V DO2 have been measured, then the system calculates the oxygen consumption according to the Fick principle. Conversely, if the oxygen consumption has been measured using the testing system, then the Fick cardiac output is calculated. Since both cardiac output and CaO2 are known, available oxygen, the product of cardiac output and arterial oxygen content, is calculated (12). We use available oxygen as the primary variable for optimizing cardiopulmonary function. This use is based on the dominance of cardiac output in peripheral oxygen delivery (13).

If end-tidal gas concentrations and mixed expired gases have been measured, then they are used for giving improved estimates of the shunt fraction, alveolar-arterial oxygen and CO2 gradients, and dead space.

Lastly, if the tidal volume has been measured in either the respiratory mechanics or the gas exchange program, then it is used for determining the dead space: tidal volume ratio.

USE OF THE SYSTEM

The blood gas entry system has been routinely used for over two years by the Pulmonary lab and has been recently introduced into the Anesthesiology lab. Entry of blood gas data has been motivated through a series of quid pro quos. Blood gas technicians are willing to enter data because the computer is used to doing all calculations automatically, to assume their data logging, report generation functions, and to transmit results directly to the ICU's, eliminating telephone calls. This solution is desirable to ICU personnel because they know that the latest results and interpretations will always be present in the computer. We have recently added the capability of notifying ICU users whenever new data have been entered into the computer by flashing a line around display screens in the ICU. This capability saves ICU users from repetitive polling of the computer to determine whether results have been entered, and is described more fully in a companion paper (14).

This system is a prototype of our goals for using an automated ICU information system. It not only records and reformats information, but it also analyzes and partially interprets. The system attempts to synthesize information through clinically based assumptions. Where assumptions are used, their use is denoted to the clinical staff. The provision of these capabilities can help to avert use of toxic FIO2 levels, and, as more information is available, can emphasize the primacy of cardiac output in protecting the PaO2.

REFERENCES

1.   WAGNER, P.D., LARAVUSO,R.B., UHL, R.R., AND WEST, J.B.:
     Continous distributions of ventilation-perfusion ratios
     in normal subjects breathing air and 100% O2. J.   Clin.
     Invest. 54: 54, 1974.

2.   GAZITUA, R., GOODFELLOW, R., VILLAR, L., WILLIAMS, L., AND
     HIRSCH., E.: An analysis of the components of the pulmo-
     nary   shunt   equation: Significance   of   the
     alveolar-arterial oxygen gradient, arterio-venous oxygen
     content difference, and mixed venous oxygen pressure. J.
     Trauma 19: 81, 1979.

3.   BLEICH, H.L.: Computer evaluation of acid-base disorders.   J.
     Clin. Invest. 48: 1689-1696, 1969.

4.   GOLDBERG, M., GREEN, S.B., MOSS, M.L., MARBACH, C.B., AND GAR-
     FINKEL, D.: Computer-based instruction and diagnosis of
     acid-base disorders: A systematic approach.   JAMA 223:
     269-275, 1973.

5.   GARDNER, R.M., CANNON, G.H., MORRIS, A.H., OLSEN, K.R., AND
     PRICE, W.G.: Computerized blood gas interpretation and
     reporting system, Computer 8: 39-45, 19 1975.

6.   THOMAS, L.J.,JR.: Algorithms for selected blood acid-base and
     blood gas calculations. J. Appl. Physiol. 33: 154,
     1972.

7.   BURROWS, B., KNUDSEN, R.J. AND KETTEL, L.J.: Respiratory In-
     sufficiency. Chicago: Year Book Med. Publ., 1975, p.
     93.

8.   KELMAN, G.R.: Digital computer subroutine for conversion of
     oxygen tension into saturation. J. Appl. Physiol. 21:
     1375, 1966.

9.   Geigy Pharmaceuticals, Scientific Tables (7th ed.).   Basel,
     Switzerland: CIBA-GEIGY Ltd., 1970, p. 617.

10. SHAPIRO, A.R., VIRGILIO, R.W., AND PETERS, R.M.:
     Interpretation of alveolar-arterial oxygen tension
     difference. Surg. Gynecol. Obstet. 144: 547-552,
     1977.

11. JANSON, C.M., BRIMM, J.E. AND PETERS, R.M.: Instrumentation and automation of an ICU respiratory testing system. Proc. First Annual International Symposium Computers in Critical Care and Pulmonary Medicine, Norwalk, Connecticut, May 24-26, 1979.

12. NUNN, J.F., AND FREEMAN, J.: Problems of oxygenation and oxygen transport during hemorrhage. Anesthesia 19: 206-216, 1964.

13. PETERS, R.M.: Life saving measures in acute ARDS. Am. J. Surgery, in press.

14. BRIMM, J.E., JANSON, C.M., PETERS, R.M., AND STERN, M.M.: Computerized ICU data management. Proc. First Annual International Symposium Computers in Critical Care and Pulmonary Medicine, Norwalk, Connecticut, May 24-26, 1979.

COMPUTER ASSISTED RESPIRATORY MANAGEMENT FOR THE CRITICALLY ILL

Judith Fulton, Judith Piggins,
Mark Halloran

Hewlett-Packard Company
175 Wyman Street
Waltham, Massachusetts 02154

Current technological advances are causing major improvements in respiratory care for the critically ill. Respiratory monitoring has developed slowly due to the lack of suitable hardware to measure pressure, flow, volume, and gas concentrations. Now increasing numbers of respiratory parameters are becoming easily available. Various techniques and equipment are employed throughout the world as hospitals reach different levels of sophistication. In some institutions ventilation management decisions are made strictly on the basis of blood gas analysis results and clinical observations. In other institutions more complex measurements are recorded and various calculations performed. The complexity of respiratory care for the critically ill patient necessitates a systematic approach to recording and organizing data. The diversity in respiratory care requires a flexibility enabling individual installations to tailor any specific form of respiratory methodology to their own needs.

The Respiratory Subsystem has been designed to provide simple, effective patient data management. It is intended for use in any intensive care setting where a significant number of patients are maintained on artificial ventilation. The Subsystem is flexible and provides the capability to draw together all available data relevant to a patient's respiratory status.

The Respiratory Subsystem exists in the larger environment of the Hewlett-Packard Patient Data Management System (PDMS). This system is a real-time multipartition software package on Hewlett-Packard mini-computers that is designed for use in hospital intensive care settings. After patient admission, the PDMS performs functions that include monitoring vital signs, calculating hemodynamic parameters, and printing patient reports.

The Respiratory package itself consists of three separate parts: configuration (performs system initialization), data entry, and data recall. It is the configuration done during system ins- tallation at hospitals that is chiefly responsible for flexibility. In addition, configuration enables new hardware to be easily incor- porated into the Subsystem.

Once the Subsystem has been configured, Respiratory Entry pro- vides a structure for nurses and lab technicians to collect data. The Respiratory Subsystem provides for both automatic acquisition of data values and manual entry of parameters not available from automatic monitoring. The physiologic values that are accumulated include arterial and venous blood gas results, hemoglobin, cardiac output, respiration rate, and tidal volume (figure 1). After data

```
MICU 2        ANDY STRAFORD            01 MAR 1444

      PHYSIOLOGIC RESPIRATORY DATA ENTRY

NO ITEM      DATA    DATE TIME    NO ITEM      DATA
-- ----      ----    ---- ----    -- ----      ----
 1.FiO2       .40    01MAR 0100   12.H CM       170.
 2.pHa       7.40    01MAR 0100   13.W KG        67.
 3.PaO2       95.    01MAR 0100   14.RESP        15.
 4.PaCO2      40.    01MAR 0100   15.IMV         14.
 5.SaO2      97.0    01MAR 0100   16.TEMP      37.1
 6.pHv       7.36    01MAR 0100   -->TV
 7.PvO2       38.    01MAR 0100   18.PeCO2       30.
 8.PvCO2      41.    01MAR 0100   19.PIP         75.
 9.SvO2      72.0    01MAR 0100   20.PEEP         7.
10.HGB       15.0    01MAR 0100   21.CPAP
11.CO         5.0    01MAR 0100


ENTER DATA (;TIME;DAY;MONTH) OR ;ITEM# GO
RELATED DATA=;77 GO              COMPUTE=;88 GO
```

Figure 1.

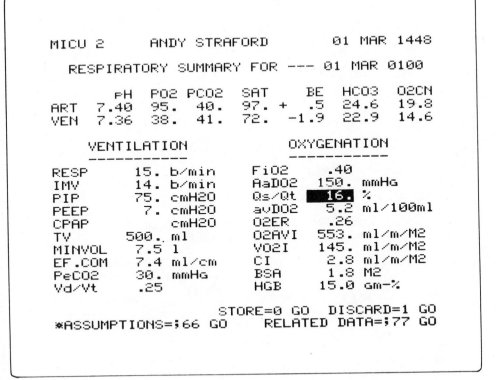

Figure 2.

is entered, the Subsystem performs calculations of derived measures of ventilatory status and performance (figure 2) and stores both entered and computed values. Approximations of ambient temperature, barometric pressure, respiratory quotient and arteriovenous oxygen difference are used as needed in calculations to provide rough estimates of derived quantities. The user specifies these assumed values during PDMS configuration (figure 3). Results computed with the approximations are flagged as estimated values.

Once information has been stored through Respiratory Entry, physicians, therapists, or nurses can request Respiratory Recall to display pertinent respiratory care and diagnosis data. The user chooses to view one of several frames that conveniently group a patient's blood gas, ventilation and oxygenation values into tabular

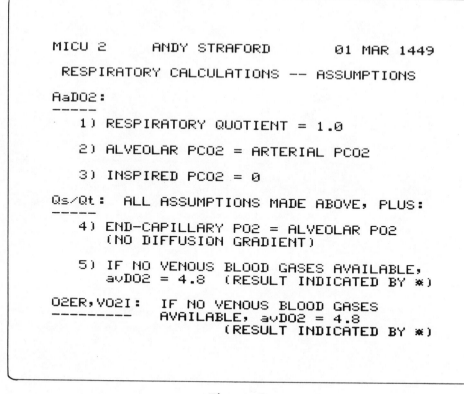

```
MICU 2        ANDY STRAFORD          01 MAR 1449

  RESPIRATORY CALCULATIONS -- ASSUMPTIONS

AaDO2:
-----
   1) RESPIRATORY QUOTIENT = 1.0

   2) ALVEOLAR PCO2 = ARTERIAL PCO2

   3) INSPIRED PCO2 = 0

Qs/Qt:  ALL ASSUMPTIONS MADE ABOVE, PLUS:
-----
   4) END-CAPILLARY PO2 = ALVEOLAR PO2
      (NO DIFFUSION GRADIENT)

   5) IF NO VENOUS BLOOD GASES AVAILABLE,
      avDO2 = 4.8  (RESULT INDICATED BY *)

O2ER,VO2I:  IF NO VENOUS BLOOD GASES
----------   AVAILABLE, avDO2 = 4.8
                  (RESULT INDICATED BY *)
```

Figure 3.

or trend plot formats. The parameters reviewed (figures 4, 5, and 6) consist of both computed and non-computed values. The calculated values fall into three categories:

1. Derived blood gas results.

2. Measures of lung behavior and effectiveness.

3. Measures of tissue oxygen requirements.

Specifically, the parameters derived from standard blood gas laboratory analysis results include base excess, bicarbonate ion concentration, oxygen content, and oxygen saturation. Relevant measures of lung behavior and effectiveness include minute volume, effective compliance, alveolar-arterial oxygen difference, shunt fraction, and dead space to tidal volume ratio. The measures of tissue oxygen requirements include arteriovenous oxygen difference, oxygen availability index, oxygen consumption index, and oxygen extraction ratio.

```
MICU 2      ANDY STRAFORD          01 MAR 1458

VENTILATION      1         2         3         4
              -----     -----     -----     -----
              01 MAR    01 MAR    01 MAR    01 MAR
              0100      0300      0530      0800

     pHa       7.40      7.40      7.40      7.40
     pHv       7.36      7.36                7.36
     PaCO2     40.       40.       40.       40.
     PvCO2     41.       41.                 41.
     RESP      15.       15.       15.       15.
     IMV       14.       14.       14.       14.
     PIP       75.       75.       75.       75.
     PEEP       7.        7.        7.        7.
     CPAP
     TV        500.      500.      500.      500.
     MINVOL     7.5       7.5       7.5       7.5
     EF.COM     7.4       7.4       7.4       7.4
     PeCO2     30.       30.       30.       30.
     Vd/Vt      .25       .25       .25       .25
                          --- NEWEST ---
                        *ASSUMPTIONS=;66 GO
     INSTRUCTIONS=; GO  DETAILED DATA=;COL# GO
```

Figure 4.

```
MICU 2        ANDY STRAFORD             01 MAR 1459

OXYGENATION      1          2          3          4
               -----      -----      -----      -----
               01 MAR     01 MAR     01 MAR     01 MAR
               0100       0300       0530       0800

FiO2             .40        .40        .40        .40
pHa             7.40       7.40       7.40       7.40
pHv             7.36       7.36                  7.36
PaO2            95.        95.        95.        95.
PvO2            38.        37.                   38.
AaDO2          150.       150.       150.       150.
Qs/Qt           16.        16.     *  17.        16.
PEEP             7.         7.         7.         7.
CPAP
avDO2           5.2        5.2                    5.2
O2ER             .26        .26     *   .24        .26
O2AVI          553.       553.       553.       553.
VO2I           145.       145.     * 134.       145.

                          --- NEWEST ---
                      *ASSUMPTIONS=;66 GO
       INSTRUCTIONS=; GO  DETAILED DATA=;COL# GO
```

Figure 5.

```
MICU 2        ANDY STRAFORD          01 MAR 1503

   RESPIRATORY SUMMARY FOR --- 01 MAR 0530

         pH   PO2  PCO2   SAT    BE   HCO3   O2CN
ART    7.40  95.   40.   97. +   .5  24.6   19.8
VEN

     VENTILATION              OXYGENATION
     -----------              -----------

RESP       15. b/min     FiO2     .40
IMV        14. b/min     AaDO2   150. mmHg
PIP        75. cmH2O    *Qs/Qt    17. %
PEEP        7. cmH2O     avDO2        ml/100ml
CPAP           cmH2O    *O2ER     .24
TV        500. ml        O2AVI   553. ml/m/M2
MINVOL     7.5 l        *VO2I    134. ml/m/M2
EF.COM     7.4 ml/cm     CI        2.8 ml/m/M2
PeCO2      30. mmHg      BSA       1.8 M2
Vd/Vt      .25           HGB      15.0 gm-%

                         RELATED DATA=;77 GO
INSTRUCTIONS=; GO      * ASSUMPTIONS=;66 GO
```

Figure 6.

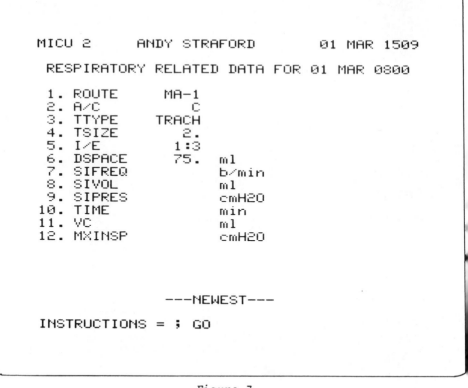

```
    MICU 2          ANDY STRAFORD          01 MAR 1509

    RESPIRATORY RELATED DATA FOR 01 MAR 0800

     1. ROUTE        MA-1
     2. A/C             C
     3. TTYPE        TRACH
     4. TSIZE          2.
     5. I/E           1:3
     6. DSPACE        75.    ml
     7. SIFREQ               b/min
     8. SIVOL                ml
     9. SIPRES              cmH2O
    10. TIME                min
    11. VC                  ml
    12. MXINSP             cmH2O



               ---NEWEST---

    INSTRUCTIONS = ; GO
```

Figure 7.

Respiratory related data may also be entered and reviewed
(figure 7) to complement the physiologic values described above.
The user is able to obtain a complete clinical setting for a particular patient by recording this additional data at the same time
as physiologic quantities. Typically the related data will include
entries that describe mechanical aspects of ventilation, such as
the type of ventilator, the type and size of airway tube, and the
inspired/expired time ratio for each breath. The respiratory related data frame is completely configurable to give each hospital
maximum flexibility in choosing what ventilation related parameters
it wishes to record.

The Respiratory Subsystem is a powerful tool for medical personnel to use in clinical settings. The respiratory data entry,
processing, and recall are oriented towards straightforward user
interaction. The Respiratory Subsystem groups data in formats
specifically tailored to health care decision making. The computer
does as much of the work as possible to minimize the time and effort required of the user, thereby freeing the medical staff to
concentrate on effective patient care.

MODULAR END-TIDAL CARBON DIOXIDE ANALYZER

H. Klement, M.S., R. Rand, M.S.,
C. Haffajee, M.D., C. L. Feldman, Sc.D.,
P. Kotilainen, Ph.D.

University of Massachusetts Medical Center
Worcester, Massachusetts

INTRODUCTION

A small, compact, modular carbon dioxide (CO2) analyzer has been developed(1) which is designed to perform as an integral part of the bedside monitoring of an intensive care patient. Designed as a plug-in module for the bedside monitor, it is convenient to use and is out of the way -- rather than adding clutter to a bedside typically overwhelmed with instrumentation. Since it is made part of the conventional instrumentation involved with ECG and intravascular pressure, monitoring its use is much more readily acceptable to nursing personnel.

DESIGN

The non-dispersive infrared analysis method of design is employed. This design principle is based on the use of a wide-band energy source and a detector that responds to only the narrow bands of the IR spectrum for CO2.

This instrument is a single beam, dual source, having a primary source and a secondary source. The primary source is the furthest away with a tungsten filament backed with a polished mirror. The energy from this electrically chopped source goes through a filter cell containing pure CO2 then the sample cell, followed by the detector. The filter cell removes the principle bands of the CO2 from the continuum radiation, so that the radiation seen by the detector is the continuum minus the principal bands of CO2. The energy from the secondary source only passes through the sample cell to the detector. This electrically chopped, servoed source is changed by the selective absorption of CO_2 in the sample cell. Since the principal bands are already removed by the filter cell, the energy from the primary source is changed only slightly by the CO_2 in the sample cell. However, the secondary source is changed to a major degree and, therefore, it is necessary to increase the intensity of the secondary source to re-establish balance. This is accomplished by servoing the secondary source so that the detector output is the same from the two sources, the magnitude of the increased voltage to the secondary source being a function of CO_2 in the sample cell.

The detector is a pneumatic, positive type, charged with CO_2. It has at its distal end a sensitive membrane of gold leaf with a stationary electrode to form a capacitor. The capacitor is polarized to 30 volts which drives an electrometer type pre-amplifier for amplification of the signal. This signal is synchronously rectified to form a D.C. error signal. The rectified signal is then integrated and the resultant D.C. error signal controls the voltage to the secondary source in a servoed way. That error voltage is also monitored by active filters to further remove the electrically chopped frequency (17 Hz) and then into a linearizer to make signal linear relative to % of CO_2 in the sample and then presented as a digital display.

Special Design Features

The inspiratory and expiratory gases are drawn from the patient to the analyzers infra-red sampling chamber through convenient and disposable airway fitting, sampling line, moisture sensor and 0.2 pore filter. These disposable sections of the sampling line are easily assembled and each part can be changed quickly and separately.

Of special interest is the moisture sensor which prevents fluid from reaching the filter and the sampling chamber where it could adversely affect the analyzer's accuracy. Once moisture bridges a small metal gap in the moisture sensor, it closes a sensing circuit which then switches a solenoid to change the route of air flow and expels the moisture back into the respiratory tubing.

The CO_2 analyzer's digital display updates every eighth breath and displays both the highest end-tidal CO_2 measured during the eight breaths, and the breathing rate. Easy to use zeroing and calibration controls are present as well as an apnea alarm which is activated after a selectable period of 17 or 34 seconds of respiratory inactivity.

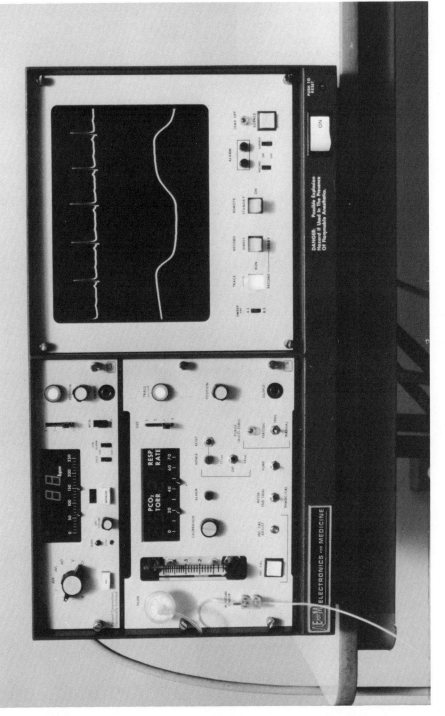

Figure 1

The analyzer outputs are accessible for long term chart recording and computer trending. The respiratory changes of CO2 concentration are displayed on the bedside monitor along with the
other physiological waveforms.

DISCUSSION

Evaluation of the effectiveness of respiratory care is usually
accomplished by taking frequent blood samples for blood gas determination. Likewise, those patients for whom respiratory ventilation is questioned, or those who have just been weaned from respiratory therapy, are also likely to have frequent blood gas evaluations. Although such samplings are necessary, it has been found
that the frequency of such tests can be significantly reduced by
the non-invasive continuous monitoring of end-tidal carbon dioxide
concentrations. The end-tidal CO2 concentrations correlate with
blood pCO2, thus this parameter is a very good index as to whether
or not the adequacy of ventilation is being maintained. A rise or
fall in the observed end-tidal CO2 concentration may indicate altered clinical status and the possible need for changed ventilation. Furthermore, it has been found that when used as a complement to blood gas sampling, the frequency of need for such samplings is often reduced.

The CO2 analyzer is a dependable warning device in the event
of failure of the ventilating mechanism, as well as physiologic
crises such as apnea. A patient's breathing rate can be seen at a
glance. If a patient starts to hyperventilate, the ICU staff is
forewarned of more serious trouble and can attempt to forestall it.
If the displayed waveform shows peaks that are sloping instead of
the normal series of plateaus, the staff is alerted that there is
maldistribution of ventilation. A patient hospitalized with severe
asthma presents a graphic example. The CO2 analyzer tracing will
have an extremely sharp slope which, as treatment progresses, will
flatten gradually to the normal plateau. The treatment's effectiveness can be gauged by the changing shape of the tracing.

The CO2 analyzer may also be of great use in surgery. Because
of its modular design, it can be easily incorporated into operating
room monitors to provide the anesthesiologist with valuable information to assure proper ventilation of patients during surgery.

CONCLUSION

The modular design of a CO2 analyzer has provided many advantages by making the continuous monitoring of end-tidal CO2 concentrations easily available. When used as a complement to blood gas sampling, the frequency of need for such samplings is often reduced. In the ICU and OR both the adequacy and distribution of a patient's ventilation can be readily assessed as well as alerting the staff with alarm conditions. An additional feature is that the disposable sampling line may be threaded into the nostril of a nasal oxygen cannula in a patient with borderline respiratory status and the adequacy of oxygenation may be indirectly assessed. This may provide a criteria for possible respirator intervention. Thus, used as a complement to blood gas evaluation for these borderline patients, and for those on a respirator, the end-tidal CO2 analyzer is a definite asset to ICU and OR monitoring.

PULMONARY ARTERY DIASTOLIC PRESSURE MEASUREMENT IMPROVED BY USE OF ANALOG CIRCUITS

P.W. Kotilainen, Ph.D., P.A. Tessier,
T.G. Stickles, and D.W. Howe, M.S.

Worcester Polytechnic Institute
Worcester, MA

INTRODUCTION

Most of the pressure amplifiers currently being used in intensive care intravascular pressure monitoring do not operate satisfactorily on patients in respiratory distress. Most pressure amplifiers using a digital display perform well on patients who are stable and have little respiratory difficulty. Depending on their basic circuitry, when used on a patient either in congestive heart failure or with other disorders affecting pulmonary function, these amplifiers will either display a diastolic pressure 5 to 10 torr low or will display a wildly fluctuating pressure. A small analog computer circuit, to be retrofitted into standard ICU pressure amplifiers, has been developed to measure and accurately display diastolic pressures.

PROBLEM

Development of digital displays on intensive care unit intravascular pressure monitoring equipment has simplified the task of viewing and recording pressures. This task was previously done by watching meter needle swings and mentally averaging extreme excursions of the needle to obtain systolic and diastolic pressures.(1) Unfortunately, the circuitry which has been used to calculate the pressure to be digitally displayed has often been inaccurate, particularly with patients who have a great degree of respiratory difficulty. Of major concern is the accurate detection and display of he pulmonary artery diastolic pressure (PAD) which is an important index of left ventricular function in lieu of obtaining frequent

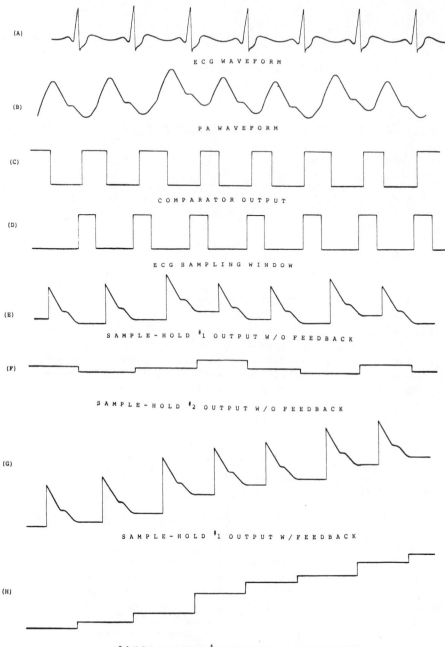

Figure 1. The accuracy of pressures determined by
needle swings averaging is also to be questioned since
due to mechanical damping of the needle, systolic pressures will be low and diastolic high.

pulmonary capillary (PC or wedge) pressures. Obviously a device which is employed to detect and display these pressures must be both accurate and stable in its detection and display.

The techniques which have been employed in the past by major medical instrument manufacturers vary in their ability to accurately detect and display a PAD being influenced by respiratory variation. The pulmonary circulation is a low pressure system. A normal pressure might typically be 25 torr systolic and 10 torr diastolic. In diseased states these pressures may increase 10 - 20 torr or more. The pulmonary pressure waveform can be modulated by 10 to 20 torr by forced respiration typical of diseased states. Since a vaiation of 5 to 10 torr can be diagnostically critical, the influences due to modulation must be eliminated.

Most medical instrument manufacturers employed some variations on the basic technique of peak detecting within a fixed or moving sampling period. It is apparent that this technique fails in the presence of respiratory variation by either peak detecting artificially low diastolic peaks or by displaying diastolic pressures which jump about dramaticallly. Whatever the final result, these amplifiers do not present a display which is useable for clinical diagnosis for those patients on whom such information is most critical.

DESIGN

To resolve the problems of inaccuracies and unreliability in PAD monitoring, a circuit was developed to account for the influence of respiration in the determination of PAD. Since respiration is periodic and varies the diastolic pressures equally above and below their true values, almost sinusoidally, it was decided that the best way to accurately determine PAD would be to average discrete diastolic pressures over a respiratory cycle. Averaging over at least one respiratory cycle would effectively cancel the effects of repiratory variation on PAD. The circuit was, therefore, designed to average discrete diastolic pressures over one or more respiratory cycles, for a specified number of heart beats.

The circuit used is essentially a special purpose analog computer which computes PAD. The computation of PAD lends itself well to digital methods which may be employed in future generations of monitoring equipment which might be microprocessor based. To solve the problem in existing equipment, which will be in use for years to come, a retro-fit of this analog circuit into standard ICU amplifiers (2) is appropriate.

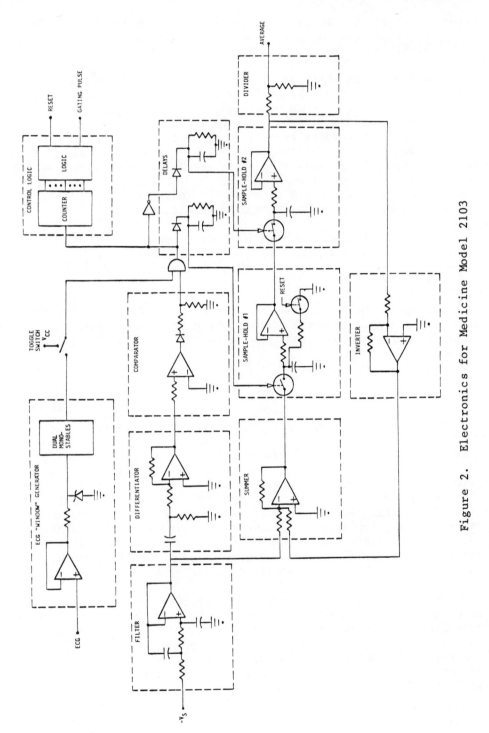

Figure 2.   Electronics for Medicine Model 2103

The input waveform of the circuit is initially filtered by a second-order, low-pass, active filter to remove artifact from the signal. The fastest heart rate to be expected would be from 120-180 beats per minute, 2-3 Hz. Therefore a cut-off frequency of 6 Hz is used to eliminate noise problems inherent to differentiation. A cut-off frequency of 6 Hz attenuates the systolic peaks while the diastolic pressures remain essentially unchanged.

The discrete diastolic pressures (minimums) are located by differentiating the filtered PA waveform. The minimums occur at the point where the differentiated signal crosses the zero axis with a positive slope. A maximum occurs when the differentiated signal crosses the zero axis with a negative slope. The zero crossing points are located by a comparator. The detection of a minimum causes the comparator output to go "high" (5 VDC). Upon detection of a maximum, the comparator output returns "low" (0 VDC).

The first Sample-Hold unit is controlled by the comparator output pulses. When the comparator detects a minimum and the output goes "high", the Sample-Hold is placed the "hold" mode. It remains in the "hold" mode until a maximum is detected and the comparator output returns "low", at which time the Sample-Hold returns to track the waveform. Thus, the first Sample-Hold stores a discrete diastolic pressure for half of a cardiac cycle.

A second Sample-Hold unit is controlled by the inverted comparator output. Therefore, while the first Sample-Hold unit is in the "hold" mode, the second unit is in the "sample" mode and vice versa. Thus, the second Sample-Hold takes the minimum from the first and holds it until another minimum is located.

In order to obtain an average of these discrete diastolic pressures, it is necessary to add a number of them together. This addition is accomplished by adding the output of the second Sample-Hold to the input signal of the first unit. The combination of these two Sample-Hold units along with the feedback loop forms an analog "accumulator".

Due to the characteristics of the comparator and Sample-Hold units, a timing problem arises in the accumulator circuitry. The problem occurs when the Sample-Hold units change state simultaneously. The Sample-Hold which is switching into the "hold" mode, tends to switch late and, therefore, misses the value which was intended to be stored. To resolve this problem, a trailing edge delay was added in the Sample-Hold control lines which come from the comparator.

The number of discrete diastolic pressures accumulated is determined by control logic circuitry, which counts the comparator output pulses. It was determined through calculations, using minimum and maximum heart and respiratory rates, that an average of ten heart beats would cover at least one respiratory cycle. Ten beats are therefore used to compute the PAD.

To obtain an average PAD, the accumulated value from the second Sample-Hold is divided by ten through voltage division. After a count of ten beats the control logic circuitry outputs a gating pulse telling the display that the average is available for its up-dating. When a count of eleven is reached, the control logic resets the accumulator so that another average can be found.

False minimums can be imposed upon the PA waveform by fluid column movements or other mechanically induced artifact. To prevent the circuitry from interpreting these minimums as actual diastolic pressures, the ECG signal is used to time a sampling "window". True diastolic pressures will occur within a prescribed time after the R-wave of the ECG signal. Two monostable multivibrators are used to form this sampling "window". The window starts 50 msec and ends 170 msec after the R-wave of the ECG signal.

This 120 msec "window" is ANDed with the comparator output to establish a dual criterion for finding a valid pressure. In order for a minimum to be considered as a valid diastolic pressure, the comparator must indicate a minimum is present and this minimum must occur withing the ECG sampling "window". The addition of this sampling window circuitry increases the accuracy of the average. However, since and ECG signal may not always be available, this portion of the circuitry is made optional through the use of a switch.

TESTING AND RESULTS

After sufficient laboratory testing using an artificial signal, a prototype circuit was retro-fitted into a monitoring unit. Upon completion of thorough electrical testing to insure total safety for the patient, the unit was clinically tested. In all tests, including those performed on several patients with extreme respiratory variations, the displayed PAD was extremely consistent on each update of the display and was shown to agree in each case with pulmonary capillary (wedge) pressure. Although the circuit operated accurately without the ECG sampling "window", accuracy was ensured with its use.

Addition of this circuit to conventional pressure monitoring instruments with digital displays, will end the present frustration of clinicians in assessing left ventricular function in those patients in respiratory distress.

INDEX

Absorption-titration technique,
 block diagram for, 84
Acid-base interpretation, in blood
 gas analysis, 383–385
Acid-base status frame, in ICU
 data base management
 system, 195
Acidosis
 mixed or metabolic, 178–180
 respiratory, 177–178
Acute postoperative circulatory
 failure, computerized
 algorithm for predicting
 outcome in, 89–100
ADB, see Amorphous data base
Adverse Trend Condition, in alarm
 systems, 292
AID interpretation language, 284
Airway disconnection, danger
 of, 173
Airway obstruction, in ventilator
 operation, 22
Airway Pressure Analysis Program,
 algorithm of, 46
Airway pressure waveform analysis,
 47
Alarm systems
 alarm types in, 293
 basic hypotheses of, 292–294
 evaluation of, 296–297
 false alarms in, see False
 alarms
 ICU setting and, 294–296
 sequence in, 295–296
 types of systems in, 293
Algorithmic diagnosis,
 limitations of, 203

Alpha ionization
 advantages of, 170
 in cardiac monitoring, 169
Alpha ionization pneumotachography
 americium as alpha source
 in, 165
 in cardiorespiratory
 monitoring, 165–170
Alpha ionization sensor, block
 circuits of, 166
Alveolar oxygen pressure,
 "ideal," 141
Alveolar ventilation
 changes in, 73
 normal pH and, 178
 $PaCO_2$ as index of, 176
Americium, in alpha ionization
 pneumotachography, 165
Amorphous data base
 defined, 197
 vs. defined data base, 197
 FREETEXT in, 199–201
 frequency distributions in,
 204
 general data type identification in, 199
 information necessary for
 clinical entries in, 198
 Main Patient Data File in,
 201–202
 MEDICATIONS in, 201
 patient data/scratch formats
 in, 200
 patient identification in,
 198–199
 Problem Oriented System and,
 201

413